The Politics of Asbestos

Pathways to Sustainability Series

This book series addresses core challenges around linking science and technology and environmental sustainability with poverty reduction and social justice. It is based on the work of the Social, Technological and Environmental Pathways to Sustainability (STEPS) Centre, a major investment of the UK Economic and Social Research Council (ESRC). The STEPS Centre brings together researchers at the Institute of Development Studies (IDS) and SPRU (Science and Technology Policy Research) at the University of Sussex with a set of partner institutions in Africa, Asia and Latin America.

Series Editors:
Melissa Leach, Ian Scoones and Andy Stirling
STEPS Centre at the University of Sussex

Editorial Advisory Board:
Steve Bass, Wiebe E. Bijker, Victor Galaz, Wenzel Geissler, Katherine Homewood, Sheila Jasanoff, Colin McInnes, Suman Sahai, Andrew Scott

Titles include:
Dynamic Sustainabilities
Technology, Environment, Social Justice
Melissa Leach, Ian Scoones and Andy Stirling

Avian Influenza
Science, Policy and Politics
Edited by Ian Scoones

Rice Biofortification
Lessons for Global Science and Development
Sally Brooks

Epidemics
Science, Governance and Social Justice
Edited by Sarah Dry and Melissa Leach

The early research for this book was done under the ESRC Science in Society Programme. Associated with this programme, Earthscan publishes the Science in Society Series, overseen by Series Editor Steve Rayner, James Martin Professor of Science and Civilization at the Institute for Science, Innovation and Society at Oxford University and Programme Director of the ESRC Science in Society Programme. Titles in this series include:

Animals as Biotechnology
Ethics, Sustainability and Critical Animal Studies
Richard Twine

Business Planning for Turbulent Times
New Methods for Applying Scenarios
Edited by Rafael Ramírez, John W. Selsky and Kees van der Heijden

Debating Climate Change
Pathways through Argument to Agreement
Elizabeth L. Malone

Democratizing Technology
Risk, Responsibility and the Regulation of Chemicals
Anne Chapman

Genomics and Society
Legal, Ethical and Social Dimensions
Edited by George Gaskell and Martin W. Bauer

Influenza and Public Health
Learning from Past Pandemics
Edited by Tamara Giles-Vernick and Susan Craddock, with Jennifer Gunn

Marginalized Reproduction
Ethnicity, Infertility and Reproductive Technologies
Lorraine Culley, Nicky Hudson and Floor van Rooij

Nanotechnology
Risk, Ethics and Law
Edited by Geoffrey Hunt and Michael Mehta

Resolving Messy Policy Problems
Handling Conflict in Environmental, Transport, Health and Ageing Policy
Steven Ney

Uncertainty in Policy Making
Values and Evidence in Complex Decisions
Michael Heazle

Unnatural Selection
The Challenges of Engineering Tomorrow's People
Edited by Peter Healey and Steve Rayner

Vaccine Anxieties
Global Science, Child Health and Society
Melissa Leach and James Fairhead

A Web of Prevention
Biological Weapons, Life Sciences and the Governance of Research
Edited by Brian Rappert and Caitríona McLeish

The Politics of Asbestos

Understandings of Risk, Disease and Protest

Linda Waldman

publishing for a sustainable future

London • Washington, DC

First published in 2011 by Earthscan

Earthscan Ltd, Dunstan House, 14a St Cross Street, London EC1N 8XA, UK
Earthscan LLC,1616 P Street, NW, Washington, DC 20036, USA
Earthscan publishes in association with the International Institute for Environment and Development

For more information on Earthscan publications, see www.earthscan.co.uk or write to earthinfo@earthscan.co.uk

ISBN: 978-1-84971 -107-4 hardback
ISBN: 978-1-84971-108-1 paperback

Typeset by JS Typesetting Ltd, Porthcawl, Mid Glamorgan
Cover design by Susanne Harris

A catalogue record for this book is available from the British Library

Library of Congress Cataloging-in-Publication Data

Waldman, Linda, 1967-
     The politics of asbestos : understandings of risk, disease, and protest / Linda Waldman.
         p. cm.
     Includes bibliographical references.
     ISBN 978-1-84971-107-4 (hardback) — ISBN 978-1-84971-108-1 (pbk.) 1. Asbestos—Health aspects. 2. Asbestos—Environmental aspects. 3. Health risk assessment. 4. Medical anthropology. I. Title.
     RA1231.A8W35 2010
     363.17'91—dc22

                                    2010032967

At Earthscan we strive to minimize our environmental impacts and carbon footprint through reducing waste, recycling and offsetting our CO_2 emissions, including those created through publication of this book. For more details of our environmental policy, see www.earthscan.co.uk.

Printed and bound in the UK by MPG Books, an ISO 14001 accredited company. The paper used is FSC certified.

In memory of Ivan Sam Waldman, 1936–2010

Contents

Preface and Acknowledgements   *xi*
Acronyms   *xiii*

1 Introduction: The Problem of Asbestos   1

2 'I've Got the Dust as Well': Asbestos Litigation, Pleural Plaques and
   Masculinity in the UK   19

3 Evaluating Science and Risk: Living with and Dying from Asbestos in
   South Africa   49

4 'Show Me the Evidence': Science and Risk in Indian Asbestos Issues   73

5 'Through No Fault of Our Own': Asbestos Diseases in South Africa
   and the UK   109

6 Reframing Risk: Comparative Framings of Asbestos and Disease   143

7 Conclusion: Diseased Identities and Social Justice   179

References   *187*
Index   *209*

Preface and Acknowledgements

Although I have authored this book, doing the research and writing has not been an isolated and solitary process. Many people have accompanied and assisted me along the way and these pages provide an opportunity to briefly document the process and my indebtedness, as well as to acknowledge diverse inputs and say thank you.

Conducting research in three different countries involved collaborations and partnerships. In South Africa, the Prieska-based grassroots movement, Concerned People Against Asbestos (CPAA), gave me open access to their material and welcomed my participation in their activities. In the UK, the Dagenham branch of Britain's General Union (GMB) allowed me to partake in their fortnightly gatherings, as well as to socialize in the pub, and attend rallies and political meetings. In India, the Society for Participatory Research in Asia (PRIA), with its long-standing expertise in occupational health, helped initiate my research, identifying potential informants and setting up interviews. No research would have been possible without this and I am deeply grateful to the CPAA, GMB and PRIA for their guidance, assistance and support.

In the UK, research centred on East London, particularly Barking (where an asbestos factory had been located) and Dagenham (where many workers exposed to asbestos continue to live). Interviews were held with key people in the Barking and Dagenham Asbestos Victim Support Group and in the Dagenham Branch of the GMB during the years 2005 and 2006. Participant observation included attending monthly drop-in sessions where people receive legal or practical advice for dealing with asbestos, fortnightly GMB meetings in the Dagenham Working Men's Club and political rallies. In addition, informal discussions with small groups of key people often developed spontaneously. This research focussed on exploring the disjuncture between people's own understandings and experiences of asbestos and disease and the ways these diseases were framed in medical and legal process. In South Africa, my research was located in the Northern Cape, where I have been conducting intermittent anthropological research for the past 20 years. Particular attention was paid to the towns of Prieska (where asbestos mines and mills had been located) and Griquatown (where residents had relocated after the mines closed). Asbestos-related research was undertaken in June and July 2003, May and June 2005, January 2007 and again in January 2008. Given my familiarity with the towns and the people, in-depth qualitative field research concentrated on asbestos sufferers' portrayal of medical and legal discourses in relation to their own personal and bodily experiences. Most of this research was conducted in Afrikaans, the lingua franca of the Northern Cape. Research in India was a new undertaking, and this was conducted primarily through open-ended, semi-structured interviews. Most of the people interviewed were fluent in English and used to dealing with western researchers, media and international organizations, although research with grassroots activists in Gujarat occasionally required translators. This research focused on exploring government policy in relation to asbestos and the process of anti-asbestos mobilization. It did not, for reasons explored later in this book, examine the experience of asbestos-related disease (ARD) and did not interview sufferers.

This book stems out of earlier research projects funded by the Economic and Social Research Council (ESRC) Science in Society Programme, the Development Research Centre on Citizenship, Participation, and Accountability and the Rockefeller Foundation. It is also inspired by the work of the Social, Technological and Environmental Pathways to Sustainability (STEPS) Centre. The opinions expressed here do not necessarily reflect those of the funding institutions. I have also received – and am grateful for – the support and intellectual stimulation from my colleagues in the STEPS Centre and the Knowledge, Technology and Society Team at the Institute of Development Studies (IDS), as well as from members of the anthropology departments of Maynooth University (Ireland) and London School of Economics (UK). I am immensely grateful for the advice, support and critical feedback received from Hayley MacGregor, Melissa Leach, Fiona Ross, Dinah Rajak, John Gaventa, Rajesh Tandon, Julia Day, Heather Williams, Nardia Simpson, Deborah James, Fiona Barbour, Cecil Scheffers and Jimmy Parish. Special thanks are due to my Claus and Tara for their constant encouragement and for putting up with me, to my mother for her endless patience and reading of drafts for grammatical consistency, and to Naomi Vernon who had the unenviable task of overseeing the editing and production of this text.

The most important people in this book, and the ones I wish to thank the most, are the residents of Griquatown, Prieska, Barking, Dagenham, New Delhi and Ahmadabad. These are the laggers, former asbestos miners, asbestos victims, activists and ordinary men and women whose stories are told within. Listening to your stories and sharing experiences of ARDs and personal distress was often a difficult and painful process for all of us. While I wish we lived in a world where such stories did not exist and where research such as this was redundant, your courage and generosity has been a constant source of inspiration to me. It is my hope that this book plays a small role in shaping a better world, in which science and technology work to help the poor and to facilitate social justice.

Acronyms

A-BAN	Asian Ban Asbestos Network
ANROAV	Asian Network for the Rights of Occupational Accident Victims
ARD	asbestos-related disease
ART	Asbestos Relief Trust
BANI	Ban Asbestos Network India
COREC	Central Office for Research Ethics Committee (UK)
COSATU	Congress of South African Trade Unions
CPAA	Concerned People Against Asbestos (Prieska)
CT scan	computed tomography scan
DGFASLI	Directorate General, Factory Advice, Service and Labour Institute
DME	Department of Minerals and Energy (South Africa)
EPH	Environmental Public Hearing
ESI	Employees' State Insurance
ESIC	Employees' State Insurance Corporation
EU	European Union
GEFCO	Griqualand Exploration and Finance Company
GMB	Britain's General Union
HSE	Health and Safety Executive
IARC	International Agency for Research on Cancer
IBAS	International Ban Asbestos Secretariat
IIAC	Industrial Injuries Advisory Council
ILO	International Labour Organization
IPCS	International Programme on Chemical Safety
MBOD	Medical Bureau for Occupational Diseases
MP	Member of Parliament
NCDRLD	National Campaign on Dust-Related Lung Diseases
NGO	non-governmental organization
NHS	National Health Service
NIOH	National Institute of Occupational Health
NIOSH	National Institute for Occupational Safety and Health
NSIL	New Sahyadri Industries Ltd
NUM	National Union of Mineworkers

NUT	National Union of Teachers
PIC	Prior Informed Consent
PIL	Public Interest Litigation
PRIA	Society for Participatory Research in Asia
STEPS	Social, Technological and Environmental Pathways to Sustainability
TB	tuberculosis
TICA	Thermal Insulation Contractors Association
UCATT	Union of Construction, Allied Trades and Technicians
UKACR	UK Association of Cancer Registries
UN	United Nations
UNEP	United Nations Environment Programme
WHA	World Health Assembly
WHO	World Health Organization

Chapter 1

Introduction:
The Problem of Asbestos

On 4 April 2007, Alasdair Packard[1] was diagnosed as having pleural plaques by a radiologist in South Africa. He had travelled from Dagenham, London, where he worked as a lagger (thermal insulation specialist), to Cape Town, South Africa, specifically for a chest X-ray. As a lagger, Alasdair had experienced massive exposure to asbestos and, as a result, he had annual medical check-ups by his local doctor in the UK to ensure that he was not suffering from an asbestos-related disease (ARD). But when these check-ups stopped performing routine chest X-rays, Alasdair was concerned: 'They always say my breathing is fine, my lungs are good, come back in two years [for your next X-ray], but I've always had that worry [of ARDs]… How are they going to find out if they aren't X-raying? You can breathe, but how do you breathe?'

South Africa has a long history of recognizing – and refusing to recognize – ARDs: pioneering medical research among South African asbestos mine workers took place in the 1950s and identified the links between mesothelioma and asbestos (Abratt et al, 2004). Despite official 'blindness' to the working conditions and health hazards of asbestos mines throughout the apartheid era (1948–1994), the government-run Medical Bureau for Occupational Diseases (MBOD) diagnosed and compensated miners for pleural plaques in a manner described by one British medical specialist in 2006 as 'generous' in comparison with the UK's provision for ARD compensation.

South African asbestos mines – many of which were owned by UK companies – flourished for about 100 years (1893–1997), producing a large proportion of the world's asbestos.[2] Much of this asbestos was exported to the UK, to the Cape plc factory in Barking, London – just a short distance from where Alasdair Packard now lives – where it was used in the production of yarn, cloth, millboard, steam packings, ropes, respirator filters, brake linings, fireproof boards and clothing as well as for insulation in large-scale infrastructure and power stations.

*When Alasdair returned from South Africa, he made an appointment
at the East Ham Chest Clinic for his usual check-up. He was informed
that he was 'fine', with his breathing improved and with 'no problems' in
his chest. To this, Alasdair responded: 'Well if you think I'm all right, I don't
think so,' and he handed over the X-rays and South African diagnosis. 'Oh
yes,' said the doctor, 'I can see that you've got pleural plaque now and you
can get your claim going,' but 'us doctors don't believe in pleural plaque.'*

*Alasdair Packard is a member of the Dagenham branch of Britain's
General Union (or GMB). Had he been diagnosed with pleural plaques
before January 2006, he would have been able to claim compensation.
But recent changes to the law have meant that he has to wait until he is
diagnosed with a more severe form of ARD before being able to take his
former employers to court. Alasdair is not alone in his struggle; he is one
of a group of laggers and trade unions who are actively challenging the
UK government's stance on ARDs. The GMB and the laggers, like many
other anti-asbestos campaigners, would like to see the UK medical and
legal systems recognize pleural plaques as a form of occupational disease
– as indeed it did between 1985 and 2005.*

This brief vignette introduces many of the issues of this book. In the first place,
it highlights the dynamics surrounding asbestos, its governance and the way
different categories of disease are recognized by different medical and legal
systems around the world. These are constantly changing domains and people
diagnosed with ARDs find themselves caught up in these dynamic situations.
As the example makes clear, medical and legal definitions of ARDs are not static
categories but are shifting in relation to governments, labour campaigns or
insurance lobbies, medical expertise and so forth. Alasdair Packard's story also
hints at the ways in which different countries – and experts within these countries
– understand and respond to ARDs. Despite the highly scientific nature of these
topics (asbestos is a geological phenomenon and ARDs are recognized medical
conditions), these varied understandings form a central theme throughout this
book. Linked to these different framings,[3] the interrelationship between science,
governance and mobilization is pursued in this book through an examination of
how people exposed to asbestos or diagnosed with a related disease understand
and interpret their asbestos-related experiences in India, the UK and South
Africa.

Asbestos once symbolized the possibilities of a new modernist era and, as I
shall later show, in India these associations still have resonance. It has also been
– and often continues to be – ubiquitous, ominous, enabling, empowering and
disempowering. As such, asbestos is a product which 'draws us in and repels
us' through its uncanny nature (Turkle, 2007, p8). An appreciation of asbestos

fibres, asbestos pollution and the associated diseases that people contract through exposure to these fibres provides insights into broader processes about society, science and technology. This involves understanding how individuals and collectivities experience exposure to toxic products, it requires an exploration of how medical and legal discourses frame disease, how people in power conceptualize asbestos and its consequences – or, put differently, how dominant narratives about asbestos exist – and it necessitates analysing processes of governance. Focusing on asbestos is also a means of opening up broader ontological questions such as the role of science and technology in society and the character of industrial growth and scientific innovation in the future.

This book is an anthropological exploration of asbestos pollution, disease and protest. It expands the study of science and society by asking questions about meaning, value and emotions. Objects, says Turkle (2007, p6), convey both ideas and emotions: 'we live our lives in the middle of things'. Asbestos, and the associated pollution and disease, have come to mean different things to different people at different times. Once used to symbolize modernization, progress and economic growth, today it is shorthand for a lack of corporate social responsibility, a lack of recognition of the victims of ARDs and a lack of governmental responsibility – but these associations are not, as this book will later show, universal. An analysis of asbestos leads one into discussions of global/local relationships and processes, of risk and uncertainty, of economic versus moral responsibility and of modern advancement in opposition to traditional stagnation. These are all themes explored in this book, which shows that such oppositional understandings do not reflect the complex, contested and often contradictory situations that people exposed to asbestos find themselves in. Using asbestos as a lens through which to focus on South Africa, India and the UK, this book explores the manner in which individual experiences of health and well-being are socially constructed and interconnected with social and political processes. It further examines the framings employed by legal, medical and political processes in relation to ARDs and the ways in which disease and identity are interconnected. There is, as Harding has argued (1996, p24), a 'good deal more thinking' to be done 'both with and about science' in order to develop specific, complete and appropriate conceptualizations of the interface between science and society which advance democratic possibilities.

The central argument of this book is that disease, compensation, identity, medical science, legal expertise, governance, social movements, politics and protest come together in different ways in South Africa, India and the UK and, in so doing, create or foreclose particular pathways to sustainability and social justice. The book identifies three potential pathways. The first involves integration of the voices of the poor through a process which exposes policy-makers and other powerful government actors to poor people's day-to-day realities and struggles in their local contexts. The second pathway employs the integration of activists and victims into science and technology debates in order to bring about the politicization of science. The third pathway entails the building of networks and alliances across political and geographic spaces, and ensuring that these networks criss-cross international, national and local contexts. Although elements of these pathways are evident in South Africa, India and the UK, the South African example demonstrates the most successful use of all three pathways. In this case, victims of ARDs have worked with non-governmental organizations (NGOs), community organizations, activists and other prominent individuals. Activists and campaigners have managed to balance grassroots initiatives and local meanings with complex international networks and alliances. This, in turn, shaped the way asbestos is dealt with within South Africa, influenced international processes of governance and supported new engagements with science in local settings.

The problem of asbestos

Asbestos is most famous for its fireproofing and insulating qualities. The word 'asbestos' stems from Greek and refers to its incombustible nature. Asbestos is a generic term for fibrous silicate minerals or, in other words, for rocks which are fibrous. It is found in several different forms, of which the most common are white asbestos (also known as chrysotile), blue asbestos (or crocidolite) and brown asbestos (less commonly referred to as amosite). The largest deposits of asbestos are to be found in Canada and Russia, but it has been mined – and in some cases continues to be mined – in Australia, Brazil, Canada, China, India, Italy, Kazakhstan, Russia, South Africa and Zimbabwe.

The fibrous nature of asbestos means that it can be woven into a thread and developed into a fabric:

> *In appearance asbestos fibres are as light as eiderdown and yet as hard as any stone. To the touch a piece of raw asbestos is like rock, but with the fingers individual fibres can be teased apart. When placed under the microscope each fibre can be seen to consist of thousands of fine threads. Asbestos can be subdivided almost infinitely until molecular dimensions are achieved. A single strand weighing less than an ounce can be spun out for three hundred feet and a square yard of woven cloth will weigh less than eight ounces.* (McCulloch, 2002, p1)

Asbestos is also very durable and long-lasting; it does not corrode easily and animals and vermin find it unattractive. It does not work as a conductor of electricity and therefore can be put to an incredible number of uses. After the two World Wars, during which asbestos had assumed immense strategic importance as a product for insulating warships, 'manufacturers promoted asbestos as a miracle product, which enhanced the quality, longevity or safety of any commodity into which it was incorporated' (McCulloch, 2002, p4). Asbestos fibre was used in a myriad of manufacturing processes including cigarette filters, mattresses, beer filters, brake linings, buildings and ships (Competition Commission, 1973; McCulloch, 2002). Asbestos, wrote Bartrip (2006, p1), 'was the perfect material for an industrialising and electrifying world of heat, combustion, and high-speed locomotion'. From this perspective, there is no problem with asbestos.

Unfortunately, however, asbestos fibres can cause a range of diseases which, alongside other toxic threats, are widely recognized for their insidious, fearsome and tainted nature (Douglas and Wildavsky, 1982; Bourke, 2005). The association of these diseases with cancer echoes a widespread connection between cancer and death (Balshem, 1991). ARDs can affect anyone exposed to microscopic asbestos fibres, although there is a raging scientific debate regarding the different types of asbestos, their toxicity and how technological processes might control exposure (discussed later in this book). In their review of ARDs, Mossman and Gee (1989) categorize four types of benign pleural disorders (namely pleural effusion or fluid on the lungs, pleural plaques, pleural thickening or fibrosis and rounded atelectasis). Most people suffering from these benign pleural disorders do not experience pain but may experience shortness of breath and some discomfort. Three forms of disease, which have more serious and debilitating consequences, are also caused by asbestos: asbestosis and lung cancer are primarily occupational hazards as contracting these diseases is linked to rates of exposure to asbestos, whereas mesothelioma (a malignant cancer) is unrelated to dosage and trivial exposure can lead to cancer of the abdominal cavity or lung lining. Mesothelioma is always fatal and people afflicted with this disease face a painful, imminent death. All ARDs have extended latency periods and only manifest themselves in physical symptoms 20–40 years after exposure to asbestos. Recent UK research on Alimta (pemetrexed disodium) has shown this to be a drug that prolongs life and alleviates the distressing symptoms of mesothelioma, but this is not a cure and all forms of ARDs are ultimately untreatable.

Estimates of how many people have been exposed to asbestos and of how many people are likely to become ill are incredibly difficult, not only because of the microscopic nature of exposure, the extended latency period and the lack of historical records detailing employment on the mines or contracting and subcontracting arrangements in industrialized countries and the difficulty (and discrepancies) inherent in diagnosis, but also because these are politically charged issues. Nonetheless, some figures are necessary to situate the discussion. One estimate, by Kasperson and Pijawka (2005), is that up to 11 million people have been exposed to asbestos in the past 70 years. The World Health Organization (WHO) suggested, in 2006, that 123 million people spread across the world still experienced occupational exposure to asbestos (WHO, 2006). LaDou argues (2004, p285), however, that 10 million people will need to die before exposure is brought to an end by a truly global ban on asbestos. The scale of ARDs, the expected rise in numbers of sufferers and the enormity of the problem have led some researchers to refer to a 'global asbestos epidemic' (Rantanen, 1997). To date, according to the International Ban Asbestos Secretariat (IBAS), a conservative estimate of work-related asbestos deaths is 100,000 people per year worldwide. In western Europe, medical experts forecast that half a million men will die from asbestos-related causes between 1995 and 2029 (Kazan-Allen, 2003).

Asbestos is only one example in an endless list of toxic wastes, industrial accidents, environmental degradation and other hazardous processes. It is, however, a hazard which has passed 'unnoticed' and 'unattended' for decades, despite the fact that its carcinogenic nature was documented in medical and legal literature from the early 1900s. It has thus been seen, as Kasperson and Kasperson (2005) argue, as a 'hidden hazard'. This characteristic of concealment stems partly from the nature of the hazard itself and partly from the nature of the society which exploits the product. The concealment of risk, in turn, is 'at once purposeful and unintentional, life-threatening and institution sustaining, systematic and incidental' (2005, p116). In addition, governments and citizens tend to concentrate on visible pollution (such as automobile fumes, smog, landfills and raw sewage) and on hazards associated with a particular place and with closely correlated cause and effect. Kasperson and Kasperson argue that factors such as the innocuous appearance of asbestos fibres, the microscopic levels of exposure necessary to cause harm, the geographic separation between extraction of asbestos and its application in industrialized settings and the extended delay between exposure and the onset of disease contribute to asbestos's elusiveness.[4] The elusiveness of asbestos is, however, also a question of framing and politics. This book thus asks to whom asbestos is elusive and hidden, who frames the fibres as innocuous and who challenges these understandings, using what kinds of framings?

All over the world, there are now people campaigning to ban asbestos and to ensure that never again will people be deliberately exposed without prior knowledge of the dangers. This has, however, particular consequences for developing[5] countries. Laurie Kazan-Allen, coordinator of the IBAS, argues that as 'industrialized nations reduced their use of asbestos, producers have increasingly targeted consumers in the developing world' (2005, p53). These processes are, as Kasperson and Kasperson point out, obscured by the political compartmentalization of the world because, as toxic hazards, they are rooted within particular values and assumptions which either elevate the benefits of such technology or underrate the consequences, and because the prioritization of economic growth over other social and political goals acts to sanction these activities. At a global level, they point out that often, in the case of 'elusive' toxic hazards, it is those who occupy the margins of societies and economies who are most affected by the hazard which 'remains concealed to those at the centre or in the mainstream' (Kasperson and Kasperson, 2005, p125).

Asbestos, its fibres and the consequences of using these products, thus permeate national and international processes such as global trade, national economic growth, national health and well-being, manufacturing standards, international trade and regulation. At a more local level, exposure to asbestos affects individuals' health, their individual identities as men and women, collective well-being, their levels of activism and desire to participate in social movements mobilizing for change as well as much of their everyday life. It is precisely because of the way discussions about the uses – and abuses – of asbestos span these different categories and contexts, that it is 'good to think with'. In order to exemplify this, this book asks fundamental questions about how asbestos is framed, by whom, and about how this affects people in India, South Africa and the UK. It explores who is affected by asbestos, how disease is defined by different actors and who benefits and who loses from these definitions, in order to demonstrate how processes of knowledge, power and political relations contribute to the 'hidden' status of asbestos.

The three countries explored in this book have all had different relationships with asbestos production, but are woven together through a history of colonialism, decolonization and development. Both South Africa and India were colonized by the UK in an era of conquest and European expansion. This was followed, in the 18th century, by a period of industrial capitalism, in which the UK invested in its colonies as it competed with other European countries for trade, resources and imperial domination. Yet, as UK companies mined asbestos in South Africa (McCulloch, 2002; Jacobs, 2003) and developed asbestos cement products in India (Tweedale, 2008), these investments were 'orientated towards imperial economic interests and needs, and institutionalized racial and cultural inequalties' (Watts, 2005, p90). In the UK itself, the era of colonial capitalism saw the importation of asbestos for the extensive manufacture and use of asbestos products. By the late 1940s and early 1950s, the management of colonies had developed into an institutional system based on technical and scientific expertise developed in the West (van Beusekom and Hodgson, 2000). Scientific innovations and increasing awareness of the dangers of asbestos were not, however, equally available to everyone, as asbestos safety procedures were recognized in the UK but flagrantly ignored in the former colonies or developing countries. Instead, as post-colonial critiques have demonstrated, replacing colonialism with modern forms of development improved the UK economy and – through perpetuating many of the 'colonial forms of power and knowledge' (Kothari, 2005, p541) – retained global inequalities between colonized and colonizing countries.

Chrysotile asbestos was banned in the UK in 1999 – crocidolite and amosite asbestos had been banned in 1985 – although asbestos exposure continues in a myriad of contexts and the country is now grappling with the consequences of its heavy manufacture and use of asbestos. In South Africa, although the mines closed in the late 1970s or early 1980s, this was not before huge numbers of people had been exposed to asbestos. The final banning of the use of asbestos in manufacturing processes in South Africa took place in 2008. In India, by contrast, asbestos manufacture is still legal. Here, the government and its citizens are only 'just beginning to consider the repercussions of the widespread use of asbestos' (Kazan-Allen, 2008, p1). The circumstances of exposure differ from country to country, as does the classification of diseases and the recognition of the dangers and risks of asbestos exposure. Yet, in all three countries people have been exposed to asbestos; they have suffered from a variety of ARDs; have mobilized to secure better recognition, compensation or legal protection; and, in all three countries, these experiences and accompanying political processes have had significant ramifications on people's individual and collective identities.

The politics of asbestos: Anthropology, science and technology studies

My first exposure to asbestos and ARDs occurred in 1998 in the Northern Cape, South Africa. At the time, I was doing fieldwork in Griquatown, working on issues of Griqua and national identity. One evening, a visiting church minister from the nearby town of Prieska explained how, in Prieska, people were queuing up to be tested for ARDs. Surprisingly, however, they were not intent on being reassured that they were healthy and that all was well. Rather, in a context of extreme poverty, they were hoping that they were suffering from a recognized ARD in order to claim the accompanying compensation. This idea – that people want to be sick and that they might desire an identity associated with disease – provides one dimension of the politics of asbestos. However, this book is not just about how ARDs and compensation might provide avenues for people to survive in South Africa. In 2005, wishing to explore more about ARDs, I found myself hanging out in the Working Men's Club in Dagenham, East London, UK. Here I met with thermal insulation engineers (or laggers) and, in speaking about their work with asbestos and their subsequent experiences, I learnt how having an ARD undermined their identities as men, as laggers and as fathers.

Here, disease erodes people's livelihoods, identities and personalities. In this context, I examine the loss of a lifestyle and the loss of well-being. In India, where I conducted research in 2007, things looked different again. Here, anti-asbestos activists struggled to secure recognition of the diseases associated with asbestos. While visiting NGO workers and activists, many of them operating despite their extremely limited resources, I came across different factors shaping people's experience of disease and identity. Here, the politics of asbestos reveals more about the roles assumed by factory managers and government officials and their confidence in technological science and manufacturing processes. It also reflects people's adoption of modernist values emphasizing economic growth, technological advancement and scientific paradigms. This, in turn, has led to a widespread lack of recognition that asbestos causes disease and a failure to associate workers' experience of illness with exposure to asbestos. In India, the politics of asbestos thus signifies a pathway characterized by denial on the part of management and government. These multiple pathways – of compensation, of identity, of denial – are not, however, as clear and delineated as they have been portrayed here and, as this book shall later demonstrate, issues such as the lack of recognition affect UK laggers just as much as they affect Indian workers – although in different ways and with different consequences. Similarly, the physical symptoms associated with ARDs undermine South Africans' attempts to establish identities and livelihoods as asbestos victims in ways that resonate with the experiences of UK laggers. ARDs also affect men and women in particular ways which reinforce stereotypical gendered roles. The politics of asbestos thus provides an in-depth exploration of the complex ways in which identity, disease, compensation, medical science, legal expertise, governance, social movements and protest come together in South Africa, India and the UK to create, or foreclose, pathways to social justice.

The approach: Technology, knowledge and identity

[T]echnology touches everyone's sense of self, all individual and collective identities. (Moore, 1996, p8)

Anthropological interest in science and technological issues is not new. Edwards et al (2007) argue that anthropology's 'refusal of domain specificity' has led to the questioning of scientific orthodoxy in relation to naturalized racial, gender and sex differences as well as in the subfields of medical and development anthropology. Anthropologists have also produced significant insights into the constructivist nature of scientific rigour and neutrality and, in so doing, have demonstrated the 'social, historical and cultural complexity of scientific fact making' (Edwards et al, 2007, p3). As a comparative discipline, anthropology has sought to identify 'scientific' attributes in other societies and in non-modern knowledge systems (Frazer, 1911; Malinowski, 1944; Evans-Pritchard, 1937; Levi-Strauss, 1963). In so doing:

> *[E]xperimental attitudes, the search for evidence, coherent argument, openness to refutation, critical thinking, doubt and operational scepticism are all found in life worlds that would not otherwise be described as either 'modern' or 'scientific'.* (Edwards et al, 2007, p3)

Anthropologists have broadened their studies beyond the scope of scientific rationality, showing how this conceptualization of knowledge overlooks emotional, ideological and symbolic dimensions associated with evidence, argument and critical thinking. The domain-specific nature of modern science and its tendency to remove knowledge from the 'social conditions of its production' has been seen as a cultural process originating in the Enlightenment period (Franklin, 1995; Edwards et al, 2007). Science, epitomized as the 'culture of no culture' (Traweek, 1988, p162), is a universalizing discourse in which scientific theory conceals, while simultaneously supporting, power relations through bounded conceptualizations of 'knowledge', 'science' and 'experts' which disempower non-specialists and non-scientists.

Anthropological conceptualizations of knowledge have nonetheless been, as Moore (1996) has argued, 'curiously divided'. On the one hand, anthropologists have been determined to recognize local knowledge as an explanatory paradigm, as a form of experimentation and as offering a means of control over nature, or as a form of 'scientific' knowledge. This recognition did not, however, extend beyond the local domain and anthropologists have been reluctant to use this conceptualization in comparative analogies that might challenge western scientific analysis. The result has been that there 'was little question of such knowledge being valorized outside the local domain' (Moore, 1996, p2). In addition to this binary conceptualization of knowledge, which both upheld local knowledge and, simultaneously and ironically, confirmed the primacy of western science, anthropologists failed to consider the politics associated with knowledge until the late 1990s. Anthropologists' attempts to deal with globalization have, however, led to the recognition that western science and technology have radically altered the contexts in which anthropologists work and, in so doing, have transformed anthropological knowledge.

Knowledge production is of course also closely associated with modern govermentality[6] as particular methods, knowledges and expertise undergird the processes of government. Specific techniques of knowledge production, delineation and rationalization allow authorities to manage, albeit remotely, individuals and collectivities (Foucault, 1991; Moore, 1996, p12). Expert knowledge and disciplinary techniques encapsulated in health care, agricultural schemes, education, development strategies and so forth are the tools of governments (Long, 1996). Individuals and collectivities are thus 'educated' through these processes of knowledge production and rationalization, and respond by regulating their personal and bodily dispositions to fit with government regimes and with dominant political and economic objectives (Long, 1996; Moore, 1996). Indeed, as Harding has argued, 'the "order of knowledge" has also been the "order of society"' (1996, p15). In seeking to understand how people in South Africa, the UK and India respond to asbestos issues, this book examines how knowledge, power and research are co-constructed and how people mobilize against powerful actors such as governments and international regimes (Jasanoff, 2004). It explores the processes by which three specific countries engage in globalized science processes and international media in order to secure particular interests (Fairhead and Leach, 2003). It examines how science serves international regimes which rely on scientifically authenticated facts and, in so doing, obscures other political and economic considerations. Indeed, as suggested by Fairhead and Leach, the 'co-production of science with policy, and the political and economic forces shaping it, has never been clearer' (2003, p2).

Knowledge and, linked to this, science and technology also shape social and political relationships (Moore, 1996) which become mediated through 'imagined communities' (Anderson, 1983) or 'imagined worlds' (Appadurai, 1990). Although enhanced by globalization and the associated rapid transmission of scientific knowledge and technology (Long, 1996, p37), this is not a one-way process in which the entire world is increasingly globalized, homogenized and 'shrunk'. Instead, processes of globalization are simultaneously accompanied by new forms of localism, by new marginalizations and by new heterogeneous compilations of identity, social networks and politics. Indeed, these simultaneously globalizing and localizing forces generate new ways of securing livelihoods, new identities, alliances and new political and social struggles, drawing on – and in the process producing – new forms of culture and knowledge (Long, 1996). Such insights have informed the ways in which anthropologists conceptualize the work of social movements and the production of knowledges as 'simultaneously local and global, but they are not universal' (Moore, 1996, p10).

Despite acknowledgement of the ways in which globalization processes reinterpret 'the local', anthropological theorizing about identity often relies on an unproblematized association with place. For this reason, Gupta and Ferguson raise the possibility of 'understanding social change and cultural transformation as situated in interconnected spaces'. This is, in effect a 'rethinking difference through connection', but always bearing in mind the ways in which global capitalism is spatialized (1992, p8). Focusing on identities as constructed in relation to local and international processes also requires a consideration of time. Identity is not a leftover remnant of ancient times. Nor is it simply a harking back for an idealized version of days gone by. Rather, as Vail (1989) argues, identity is Janus-faced, it is a collective consideration and reflection of the past but it is also about constructing a viable future. Within this book, various different kinds of identities are considered, including individual, collective, ethnic, class and gendered identities. In addition, and as this book will later show, new identities cutting across national and international 'spaces' are shaped as people mobilize against proscriptive legislation. These identities are multiple, fluid, shifting and, at times, contradictory. Thus, as S. Hall argues, identity is 'a kind of unsettled space ... between a number of intersecting discourses' (1989, p10). Finally, in thinking about identities, this book draws on actor-orientated approaches exploring people's mobilization around asbestos issues, looking at what it means to them to be part of a network and how connections between people shape their identities. As demonstrated in later chapters, NGOs, community organizations, activists and other prominent individuals have sought to establish complex international networks and alliances in order to shape the ways in which various countries deal with asbestos issues. The book demonstrates where these 'voices from below' have been able to influence international processes and governance (Leach et al, 2005) and how social movements and citizen action have supported new engagements with science in local settings (Waldman and Maat, 2007).

Making science and technology work for the marginalized

Harding has argued that 'the natural and social sciences we have are in important respects incapable of producing the kinds of knowledge that are needed for sustainable human life in sustainable environments under democratic conditions' (1996, p15). Because social and natural sciences were developed in order to administrate and manage nation-states, militaries and international systems of production, their methods and concepts of rationality and objectivity need to be reclaimed, re-envisaged and reformulated in order to achieve more democratic and sustainable ambitions. This involves individuals, other collectives such as communities, organizations, ethnic groups or social movements, and societies understanding when to use one kind of science and when to rely on another form of science and evaluating modern knowledge in relation to other knowledge systems. As such, instead of a scientific 'perfect' or 'accurate' representation of the world, she advocates searching for 'useful' knowledge and building up collages which facilitate the application of different sets of knowledge to different problems. Increasingly, it has been recognized that the role of science, in so far as it extends beyond laboratory experimentation, is 'visibly conducted, contested and authenticated' in processes involving legislation, medicalization and policy. Indeed, as Fairhead and Leach have argued, the 'conduct of science is becoming important to moral and economic contestation' (2003, p2). The practice and contestation of science thus shapes many aspects of social life including, as suggested above, mobilization, identity, gender and political authority.

This book is one in a series of books produced by the Social, Technological and Environmental Pathways to Sustainability (STEPS) Centre. It is an exploration of the ways in which social, technological, environmental, governance and knowledge processes around asbestos interact. The book adopts a broad understanding of sustainability: as a dynamic and political concept which refers to and integrates 'specific qualities of human well-being, social equity and environmental integrity' (Leach et al, 2010, p5) while simultaneously paying attention to the processes required to maintain well-being, equity and environmental integrity. This means that conceptualizations of sustainability are almost always contextual, contested and politicized – as different categories of people mobilize and negotiate to make their versions of what must be sustained 'stick' in national policy. Thus, in this book, sustainability refers to the ways in which different people (activists, men, women, asbestos-disease sufferers, scientists, politicians, business representatives and so on) understand asbestos, its associated diseases and its impact on the environment, and seek in turn to inject their values and definitions into political processes, shaping what should be done in order to achieve sustainability within asbestos-polluted environments and for diseased populations. As such, this book addresses the challenge of building pathways to sustainability and social justice through exploring the particular issue of asbestos exposure and disease in a highly complex and dynamic world. In the chapters that follow, I examine how, despite the widespread recognition that asbestos is carcinogenic, regulatory structures use different scientific and medical framings to assess risk, determine appropriate legislation, gauge people's suffering and set commensurate compensation. Official framings of asbestos risk tend to exaggerate the formulaic, scientific calculation of likelihoods and outcomes, while downplaying situations where knowledge about outcomes and likelihood are not as clearly defined (Stirling, 2009). Thus, conditions of uncertainty, ambiguity and ignorance are effectively excluded from governance, policy and other procedural processes (Scoones et al, 2007). An indication of these more intangible ideas, which challenge dominant framings of risk, is to be found in workers', sufferers' and activists' understandings of ARDs, danger and appropriate regulation. For example, as demonstrated in Chapter 4, the Indian government and private businesses construct chrysotile asbestos as safe and hence as a controllable risk in certain environments. This dominant narrative is supported through the parameters set in research projects, through legal and medical framings of ARDs and through the formal processes of documenting rates of ARDs. However, these official notions of risk are juxtaposed with alternative understandings, or narratives, about chrysotile as carcinogenic and uncontrollable. Linked to these various and often contradictory narratives regarding how to understand asbestos are different possibilities of action. In addition to examining the more formal, dominant narratives of asbestos risk, this book focuses on conditions of uncertainty, ambiguity and ignorance as constructed in alternative, non-dominant framings of asbestos and its consequences. In South Africa, India and the UK, socio-cultural, political and economic conditions shape people's responses to asbestos in different ways. These alternative narratives and framings challenge ·and contradict the various countries' biomedical and/or legal understandings of ARDs producing different identities and diverse pathways to social justice and sustainability.

Focusing on narratives, framings and people's experiences, this book represents an anthropological exploration of science and society. In this sense, the book is an 'ethnography of the particular' which aims to challenge processes of 'othering' and to 'write against culture' (Abu-Lughod, 1991). Having three different research sites enables me to discern the different ways that disease and identities are experienced in different contexts. This is not to simply state the differences but rather, following Marcus's call for a multi-sited ethnography that deliberately seeks to cross national boundaries, to explore the 'nature of relationships between sites of activity and social locations that are disjunctive, in space or time, and perhaps in terms of social category as well' (1999, p6). This book explores interaction between located meanings and global processes examining sites which appear completely separate from one another, but that have been – and continue to be – interconnected through global political and economic processes. Focusing on both northern and southern contexts, it investigates solidarities and identities that are both 'physically grounded in place' and which relate to – and perhaps rely on – internationalized, deterritorialized networks and negotiations. As such, the research recognizes Gupta and Ferguson's argument that the 'partial erosion of spatially bounded social worlds and the growing role of the imagination of places from a distance ... must be situated within the highly spatialized terms of a global capitalist economy' (1992, p11). This examination of the political economy of asbestos production and disease, through dominant and alternative narratives, explores the way knowledge and framings make invisible the connections and exercise of power.

This multi-sited research undermines north/south dichotomies. In addition to the choice of field sites, which acts as a recognition that cultural differences are located 'at home' as well as in 'exotic' societies (Marcus and Fischer,

1986), this book juxtaposes the literatures emerging from these different sites, including northern-based science and technology studies and more southern-focused anthropological work. This book thus aims to 'stress the importance of knowledge production at the margins, of being in exile from the centre; and this decentering of the subject' (Moore, 1996, p9), while recognizing that these margins are to be found in northern and southern contexts. This involves examining how knowledge works in different places, and how different people's positionality affects their knowledge with the aim of demystifying 'science through the ethnographic study of scientific practice and everyday knowledge' (Long, 1996, p57). The anthropological basis of this book is complemented by an actor-orientated approach that seeks to understand the construction of knowledge in relation to personal illness, medical science and litigation, and that focuses on the practices of everyday life, on people's strategies, mobilization and struggles regarding social identity in order to appreciate 'encounters between different kinds of knowledge and ideology' (Long, 1996, p57).

Outline of *The Politics of Asbestos*

Chapters 2, 3 and 4 focus on country-specific analyses, using an expanded case-study approach. Chapter 2 is UK-focused and explores the asbestos-related experiences of laggers (or thermal insulation engineers) in the UK. The chapter examines medical and legal framings of ARDs in the UK and complements these formal framings with an investigation of laggers' personal and bodily experience of ARDs. These emic[7] perspectives are then contextualized through examination of the social and political-economic ramifications of disease and litigation in a context of increasing distrust of science and of authority. Chapter 2 argues that victims of ARDs draw on – and extend – medical and statistical estimations of disease, but also challenge medical notions of cause and effect while advancing their perspective based on lived experience. Ultimately, the chapter shows that laggers are facing uncertain futures as a result of industrial exposure to asbestos, while their way of life and socio-economic standing is also under threat. Chapter 2 is also about gender and, in particular, about laggers' masculine identity that – although bolstered through their profession in the past – is now being eroded by a combination of socio-political conditions and by their contraction of ARDs.

Chapter 3 provides a particularly South African perspective of ARDs, focusing on impoverished former asbestos mine workers and residents in the Northern Cape towns of Prieska and Griquatown. In keeping with the previous chapter's focus on people's own understandings of ARDs and exposure, it explores town residents' and former mine workers' interpretations of health, illness and well-being. The chapter argues that widespread asbestos pollution and the risk of contracting an ARD give rise to medical, scientific discourses about the probability of technical risks which coexist with local cultural perceptions of danger and exposure. Scientists (and associated bureaucracy) and communities differently construct reality around asbestos pollution and its dangers, and identify different sources of danger. Chapter 3 ultimately demonstrates that, because understanding scientific knowledge is linked to activities and is skill-based, community participation can result in a broadened understanding of risk that draws on both scientific understandings and community perspectives of risk.

Keeping the focus on southern examples, Chapter 4 examines the dynamics around the construction of asbestos risk and disease in India. Here, in contrast to the South African example, science and medical opinions regarding asbestos are the preserve of the elite and expert opinion works systematically to exclude local understandings of disease and risk while shaping government policy. This chapter explores the political economy of asbestos in India, focusing on the supposed absence of ARDs. It shows how asbestos has been considered as a tool for Indian economic growth and modernization and how the scientific debates have constructed its 'safe' use through framings. Chapter 4 argues that the state's narrow definition of ARDs enables it to document officially the lack of ARDs experienced by Indian workers. This process, which defines sufferers as politically invisible and inconsequential, accompanied by the 30-year delay between exposure and the onset of disease, hinders anti-asbestos organizations as there is no constituency to be mobilized.

Chapters 2, 3 and 4 demonstrate similar themes being played out in very different contexts. All three chapters are about how people are defined as asbestos victims and about the interaction – shaped around these definitions – between them and various forms of governance. In all these chapters, the focus is on 'diseased identities' of one sort or another. While the UK-based case study shows how ARDs undermined men's masculine identity, the South African example provides an exploration of how having the disease facilitates community participation and reshapes scientific knowledge for both scientists and town residents. In this way, ARD sufferers and other people exposed to asbestos demonstrate an identity shift from victim to active participant in policy and, in so doing, shape not only their own knowledges but also those held by experts. In the Indian case, the identities are notable not by their active mobilized strategies, but rather by their absence. It is through processes of governance, informed by scientific, medical and legal understandings of knowledge, that asbestos-disease sufferers are denied any related form of identity.

Chapters 5 and 6 are more explicitly comparative and focus on thematic interests rather than geographic contexts. Chapter 5 reintroduces the theme of gender in relation to diseased identities and focuses on a South Africa/

UK comparison. It builds on the earlier case study chapters by contrasting authoritative and emic values through the examination of governmental recognition of risk and people's own understandings of danger. The chapter also broadens the earlier focus on masculine identity by exploring both men and women's diseased identities and contrasting experiences in the UK and South Africa. Using an 'effects made by gender approach', the chapter examines how science intersects with gender, leading people to emphasize conventional gender roles as their identities. The chapter contrasts people's 'subjective' interpretations with the country-specific medical, legal and political categorizations of disease with which these people regularly engage – highlighting how people experience, interpret and respond to ARDs.

Chapter 6 is a three-country comparison that explores the different ways in which country-specific framings incorporate social, environmental and technological dimensions. This links the idea of elusive hazards to framings and shows how different framings underplay or eliminate the understanding of asbestos as a potential problem. In order to reveal different framings, the chapter includes an exploration of ARD narratives, exploring which are 'dominant' or 'alternative' in each of the three countries, summarizing which actors are producing these narratives, what kinds of knowledge/identity they are drawing on, which social, technological or environmental dynamics are highlighted and how issues of uncertainty are dealt with.

In conclusion, Chapter 7 focuses on the marginality and identity of people exposed to asbestos and experiencing ARDs. Through an exploration of pathways, it links the people discussed in the book to the larger collectivities of the nation-state as well as to the broader STEPS themes of social justice and pathways to sustainability.

Notes

1   In accordance with the UK National Health Service's (NHS) Central Office for Research Ethics Committee's (COREC) requirements, pseudonyms have been used for all participants in this research, with rare exceptions when individuals' asbestos activism has already been documented in publications.

2   According to the South African Minister of Environmental Affairs and Tourism, South Africa 'produced 97% of the world's crocidolite, 100% of the amosite and was the 5th largest producer of chrysotile' (2008).

3   Framing refers to the ways in which scientific topics and policy processes are delineated. Cultural contingencies, life experiences, intellectual paradigms and political agendas are often highly influential in shaping how science or policy is conceptualized. These unrecognized 'blinkers' limit the possibilities to recognize multiple perspectives, seek more participatory solutions or question the assumptions

on which decisions get made. Leach et al thus argue for the necessity of recognizing other kinds of knowledge shaped – or framed – through 'other practical cultural assumptions, meanings and life-worlds' (2007, p8).

4   Kasperson and Kasperson (2005) include multifaceted human and natural physical interactions and the protracted accretion of toxic materials in environmental sinks as characteristics of this type of hazard.

5   A wide range of terms are used to characterize societies today. Often expressed as binary opposites these include undeveloped/developed, underdeveloped/overdeveloped, modernized/traditional, first world/third world, north/south, industrialized/less industrialized. No single term captures the complexity of changing economic, political, historical and cultural processes in any of these societies. In this book, I have chosen to use the terms 'developed' and 'developing' in order to emphasize the complex, unequal political and economic relationships which create these two extremes. There are, nonetheless, limitations in choosing these terms which overlook cultural understandings of development and which prioritize economic measures.

6   'Governmentality' refers here to the art and practices, or 'technologies', of governments. It refers to the manner in which problems are conceptualized by authorities and how these conceptualizations shape particular interventions or processes. It is thus about the 'specific discourses and practices, and with the particular rationalities which sustain them in the context of a given set of material and historical conditions' (Moore, 1996, p12; see also Foucault, 1991).

7   'Emic' refers to people's own interpretations or to behaviour explained in terms meaningful to the actor.

Chapter 2

'I've Got the Dust as Well': Asbestos Litigation, Pleural Plaques and Masculinity in the UK

Introduction

In the UK, in October 2007, the House of Lords unanimously ruled that sufferers of the asbestos-related disease (ARD) pleural plaques could no longer be compensated in English law. The Law Lords' position, as stated by Lord Hope, was that pleural plaques are symptomless, cannot be seen or felt, do not lead to other harmful experiences and have no consequences. As such, this disease cannot attract an award for damages which are given for 'injuries that cause harm, not for injuries that are harmless' (*Grieves and others v F. T. Everard & Sons and others* [2006] UKHL 39, paragraph 47). In the words of one protester, the Law Lords had declared that people suffering from pleural plaques had 'sustained damage, but had not been harmed'. This legal stance affirms the decision of the Court of Appeal of England and Wales, taken in January 2006, that pleural plaques are 'inert' and therefore not subject to compensation by law (*Rothwell v Chemical Insulating Co. Ltd and another* [2006] EWCA Civ 27). According to these legal assessments, pleural plaques do not involve physical symptoms or bodily deterioration and are not the precursor to more debilitating ARDs such as asbestosis, lung cancer or mesothelioma. These decisions, by the House of Lords and by the Court of Appeal, reversed legal precedent as litigation for damages resulting from pleural plaques had occurred regularly between 1984 and 2005. Indeed, legislation aimed specifically at pleural plaques was introduced in 1985, making it possible for claimants to opt for provisional damages. This meant that people diagnosed with pleural plaques could 'make a claim on the evidence available at one point and return to court later if their condition deteriorates or new asbestos-related disorders emerge' (Felstiner and Dingwall, 1988, p11). This provision proved too unpopular and many claimants pursued full and final settlements when diagnosed with pleural plaques. Such a choice was contrary to medical advice and inexplicable to the judges involved in the High Court and Court of Appeal hearings.

This chapter uses laggers' (or thermal insulation engineers') experience of pleural plaques to explore the current framing of pleural plaques and other ARDs in the UK. It examines recent court cases and victims' experiences of ARDs and shows how different the medical, legal and emic perspectives of disease can be. The chapter goes on to explore the social and political-economic ramifications of ARDs and litigation in a context of increasing distrust of science and of authority (Adam et al, 2000; Beck, 1992; Furedi, 1997/2002). It argues that victims of ARDs draw on – and extend – medical and statistical estimations of disease, but also challenge medical notions of cause and effect while advancing their emic perspective based on lived experience. Thus, the idea that pleural plaques are benign is completely absurd from the perspective of those who experience the disease and for whom the very thought of ARDs is linked with death.

The legal categorizations used in the House of Lords and Court of Appeal are based on medical assessments of ARDs and bodily harm. In the UK, the following ARDs are generally recognized: mesothelioma, lung cancer, asbestosis, pleural thickening and pleural plaques. Most medical research and litigation focuses on mesothelioma which is widely recognized as the most serious and debilitating ARD. As one doctor succinctly summarized: '[Life expectancy is] nought to two years with zero recovery, regardless of treatment.' Pleural plaques, on the other hand, are seen as something, in the words of one UK lawyer, 'you'll die with it, not of it'. These assessments of ARDs – that mesothelioma is the most debilitating and that pleural plaques are benign – are made by medical experts and, as illustrated in January 2006 and October 2007, influence the legal categorization of harm.[1] Dangers of unwarranted compensation and increased stress were listed among the primary reasons for these decisions: 'There is a danger that those ... who make a business out of litigation, will encourage workers who have been exposed to asbestos to have a CT scan in order to see whether they have pleural plaques for the sole purpose of bringing claims for compensation. Such a practice will tend to create stress and anxiety where none exists' (*Rothwell*, paragraph 67 per C. J. Phillips and L. J. Longmore). Claimants, however, take a different view and, as their decision to opt for a full and final settlement for pleural plaque shows, do not necessarily support medical and legal assessments of their physical condition.

Working with asbestos

In 1998, 12 members of Britain's General Union (GMB) in Dagenham were diagnosed with pleural plaques. These laggers have worked with asbestos virtually all their adult lives and have intimate experience with ARDs. Their earning capacity and their identity as men are intricately related to their experiences as laggers in Dagenham, East London. Dagenham is part of the London Borough of Barking and Dagenham located on the river Thames. In its heyday, Barking and Dagenham was highly cosmopolitan with companies such as the Ford Motor Company initiating new mass production methods and providing employment opportunities. Shipbuilding, electricity generation and aircraft, chemical and motor manufacturing were all popular industries between the 1930s and 1960s and all relied heavily on asbestos. Workers in the construction and transport industries were attracted to the area. Today the area remains heavily industrialized, more so than any other London borough. This, accompanied by the decline in the manufacturing sector, has meant that it is no longer an area of opportunity. It is now criss-crossed with motorways and covered with sprawling council estates. Barking and Dagenham borough has the lowest average income in London with most people earning in the region of £13,000 a year, accompanied by low levels of education. In addition to these structural conditions, 'there are high levels of long term illness and men have the third lowest life expectancy in London' (Barking and Dagenham Council, 2002, p4).

Asbestos was first brought into the UK in 1857 with the first processing plants being set up in the 1880s. In 1929, Barking Council initiated its first inquiry into the dangers associated with ARDs. By 1930, Members of Parliament were informed of an 'epidemic of asbestos disease among British asbestos workers' and this resulted in the 1931 Asbestos Industry Regulations. These regulations determined a 'safe' level, calculating that a worker who had between 15 and 20 years exposure had a 33 per cent chance of contracting asbestosis.[2] These 'safe' levels were amended in 1960 which increased the legal limit of exposure and as a result enhanced workers' chances of contracting ARDs. In 1968 standards were adjusted downwards, and were later estimated to reduce risk of asbestosis to 10 per cent (London Hazards Centre 1995). Since the 1980s, official policy has prohibited trade, application and supply of blue asbestos (crocidolite) and brown asbestos (amosite). A host of legislation has been enacted to control work environments and asbestos-related risks including the Health and Safety at Work Act 1974, the Asbestos (Licencing) Regulations 1983, the Control of Asbestos at Work Regulations 1987, amended in 1992, the Control of Asbestos in the Air Regulations 1990, the Asbestos (Prohibitions) Regulations 1992, the Control of Asbestos at Work Regulations 2002 and the Control of Asbestos Regulations 2006. Generally speaking, asbestos is considered to have been banned in 1985/1986 (Kazan-Allen, 1999; London Hazards Centre, 1995; Gee and Greenberg, 2002), although this banning was restricted to amosite and crocidolite. Chysotile asbestos was only banned in 1999, through the Asbestos (Prohibitions) (Amendment) Regulations 1999 (Kazan-Allen, 1999).

The laggers have lived in Barking and Dagenham all their lives. The oldest of the GMB laggers interviewed began work in 1961, just when exposure levels were legally increased and continued to work with asbestos until it was formally banned. Since then, many report involuntary exposure to asbestos, believing that the products or sites they were working with and on were safe and only subsequently finding out about the presence of asbestos.

The official medical and legal views

Scientific medicine, or biomedicine, and western doctors' positions and knowledge are upheld in the UK's legal system. As such, it is presented as a system of knowledge that reflects and describes natural, biological order and disorder. Medical anthropology has, however, argued that medical science is in fact part of a cultural system, a 'highly specialized version of reality' that, instead of mirroring nature, constructs bodies as biological objects which exhibit measurable 'signs' or physiological irregularities (Good, 1994, pp5, 8; see also Kleinman, 1974, 1977, 1981). People's conditions can thus only be considered 'real' if they reflect medically recognized physiological conditions which can be empirically documented. Nonetheless, as Helman (1984) points out, medical doctors exercise substantial powers over their patients, questioning them, examining them, prescribing drugs and so forth, based on the doctors' expertise and medical knowledge. More significantly:

> [T]hey can also label their patients (sometimes permanently) as ill, incurable, malingering, hypochondriacal, or as 'fully recovered' – a label which may conflict with the patient's perspective. These labels can have important effects, both social (confirming the patient in the sick role) and economic (influencing health insurance or pension payments). (Helman, 1984, p51)

The power to label patients is reinforced by the way the legal system prioritizes doctor's diagnoses and the opinions of medical experts. In the case of ARDs, this is particularly pronounced as there are various forms of compensation associated with the different ARDs. Pleural thickening, asbestosis, lung cancer and mesothelioma are all diseases for which legal compensation can be sought, but pleural plaque disorders are not. This is in keeping with Dr Rudd's perspective. As a leading UK expert in respiratory medicine, his work has been invoked in legal battles to bolster the argument that pleural plaques should be excluded from the legal framework of ARDs:

> *Pleural plaques are not thought to lead directly to any of the other benign varieties of asbestos-induced pleural disease, nor to pose any risk of malignant change leading to mesothelioma. Their presence may indicate, nevertheless, a cumulative level of asbestos exposure at which there is an increased risk of mesothelioma or other asbestos-related disorders. On average, in the absence of any other evidence about exposure it is reasonable to assume that subjects with plaques will have had higher exposure to asbestos than subjects without plaques. The frequency of development of other complications of asbestos exposure in persons with plaques is not a function of the presence of the plaques, but of the asbestos exposure that caused plaques. Since plaques may occur after a wide range of different exposures, the risks of other asbestos-related conditions may differ widely between different populations and individuals with plaques.* (Rudd, 2002, p344)

Historically, compensation was awarded for pleural plaques in the UK. In the mid-1980s, the High Court of England and Wales ruled that this disease could be seen as an 'actionable injury' because changes in the structure of the pleura were, in effect, a form of damage, because of the risk of further disease and because of the associated anxiety (Steinberg, 2006). At this stage, pleural plaques were considered to be one of a range of ARDs, less serious than many others, but nonetheless the one that was most frequently litigated. In the following 20 years, the amount awarded in damages rose significantly from £1250 in 1986 (as a provisional damage) to between £12,500 and £20,000 as full-and-final settlements. This was, in many respects, advantageous to the laggers. For example, in 1988, Edward Freeman was diagnosed with pleural plaques and emphysema. On commencing an action for damages, he was expecting about £12,000 in compensation, so when he was offered £22,500 as an out of court settlement, he accepted immediately. When interviewed in February 2006, Edward Freeman had experienced no further symptoms and considered himself lucky.

Faced with an environment in which claimants had increased access to courts and a rising trend in asbestos litigation (Felstiner and Dingwall, 1988), insurers and companies expressed concern regarding escalating payments. They argued that the 'nationwide litigation of asbestos-related diseases, and *pleural plaques in particular*, was financially ruinous' (Steinberg, 2006, p5, my emphasis) and asked for an investigation into ARDs and their actionability. This investigation took place in the High Court of England and Wales in 2005 through the test cases of *Grieves and others v F. T. Everard & Sons and British Uralite Plc and others* [2005] EWHC 88 (QB) (note that the appeal of this group of cases is *Rothwell*). The defendants argued that pleural plaques caused only minimal damage, did not result in respiratory complications or pain and were merely indicators of exposure to asbestos as 'the presence of plaques does not in itself give rise to an increased risk of malignancy'. In terms of the anxiety, they argued that first, this was related to the exposure to asbestos, rather than to the identification of pleural plaques on their lungs and that it was the anxiety which led them to undergo medical investigations for pleural plaques in the first place. Second, given the low level of risk involved, they argued that this anxiety was 'irrational'. Thus, as there was no 'real' physical damage, the claimants' anxiety was insufficient grounds for compensation (Steinberg, 2006, p6). The claimants argued, however, that a breach of duty had occurred, that damage need not produce symptoms, pain or impairment to be real, that 'irreversible structural damage to the architecture of the lungs' had occurred which, in conjunction with the associated risks and anxiety was more than minimal (Steinberg, 2006, pp7–8). The judge, Justice Holland, based his case on the previous House of Lords' decision in *Cartledge v Jopling* [1963] AC 758 and concluded the permanent ingestion of asbestos fibres to be the real problem. He ruled, on 15 February 2005, that damages could be awarded where there was an *assessable risk* of further disease. Thus, while the claimants won this case and confirmed the actionability of pleural plaque, Justice Holland emphasized the need for 'moderation' and downgraded provisional damages for pleural plaque to £3500–4000 and full and final settlements to £6000–7000 (*Grieves and others*, paragraph 94).

Although this was a partial victory for the claimants, the case was appealed by the insurers who felt that they were not liable to compensate pleural plaques at all, or alternatively that the damages were still significantly higher than what had been paid in the mid-1980s. At the appeal, both parties argued that *Cartledge* set a precedent in their favour. The defendants stressed the implications of the case in terms of policy: among other things, they were concerned that Justice Holland's decision would lead to wide-scale litigation for the fear of disease, the additional expenses of insurance liability and potential conservatism when testing new drugs. Indeed Justice Holland himself expressed concern that the actionability of pleural plaques would signify to claimants that their anxieties were justified, despite doctors' reassurance to the contrary. In addition, he argued that the legal process, in itself, created anxiety. The claimants argued that a historical precedent had been set for the actionability of pleural plaques, including decisions by three High Court judges, and that the policy implications were irrelevant. Other considerations included that changing standard legal practice after almost 20 years would bring the law into disrepute, that it would result in an increase in litigation concerning the nature of medical diagnoses, increased cases examining claimants' anxiety regarding pleural plaques as recognized psychiatric responses and increased medical testing for asbestosis (Steinberg, 2006, pp11–14). The Court of Appeal recognized that exposure to asbestos had three consequences: pleural plaques; a risk of developing more serious ARDs;

and anxiety linked to developing these more serious conditions; but noted also that taken independently, each was insufficient grounds for litigation. Although UK law had allowed these three consequences to be aggregated, there was no legal precedent or, it argued, logical reason for doing so.

Removing pleural plaques from the litigation process has been seen as enormously beneficial to insurers who estimated that approximately £1–1.4 billion had been saved (Eversheds Briefing, 2006). The result is that the possibilities for laggers – and other claimants – to sue for ARDs are considerably more limited with significant ramifications for how laggers experience the medical and legal systems and for how they live their lives once diagnosed (discussed in more detail below). The diseases for which people can now legally claim compensation have serious and debilitating consequences. Asbestosis and lung cancer are primarily occupational hazards, as contracting these diseases is linked to rates of exposure to asbestos. Although the risk of contracting mesothelioma (a malignant cancer) is related to dosage, even trivial exposure can lead to cancer of the abdominal cavity or lung lining.

Medical definitions point out that asbestosis is primarily an interstitial lung disease affecting the 'internal' tissue of the lung where gas exchange occurs, whereas pleural plaques are situated on the lining of the lung, or the parietal pleura and diaphragm (Rudd, 2004). For a respiratory physician, the two diseases have very different pathologies. These distinctions do not, however, always survive in legal and other descriptions of the diseases. Asbestosis, according to one law firm, is 'scarring of the lungs, which in severe cases causes extreme breathlessness' making patients susceptible to pneumonia and other conditions (Field Fisher Waterhouse, n.d., p3). The National Health Service (NHS) describes the symptoms of asbestosis as breathlessness (enhanced through exercise), coughing, chest pain and a constriction of the chest. It continues:

> *Asbestosis may damage the function of the lungs so much that the condition progresses to respiratory (breathing) failure. At this stage the oxygen supply to the body is so poor that the patient is always breathless and has blue-tinged skin (cyanosis) even when at rest in bed.*[3]

The International Ban Asbestos Secretariat (IBAS) argues that: 'In asbestosis, lung tissue is scarred and thickened by the action of the asbestos fibres in the alveoli, the air sacks.'[4] These accounts, with their emphasis on scarring of the lungs, are in many respects remarkably similar to that of pleural plaques, although pleural plaques may not be litigated for. Indeed, Steinberg argues that the 'pathogenesis [of pleural plaques and diffuse pleural thickening] is thought to have more similarities than differences' (2006, p1).

The similarity between the definitions of asbestosis and pleural plaques raises, for many sufferers, questions about medical definitions and legal categorization of compensation. In 2000, Wikeley argued that, in Scotland, the absence of any 'clear statutory guidance' meant that medical practitioners were exercising individual judgements when making their diagnoses. He further suggested that asbestosis was a term used to cover a 'range of conditions of varying severity', which allowed doctors to make more or less conservative diagnoses, based on 'difficult questions of judgement' (2000, p116). Felstiner and Dingwall recognized similar processes in England where medical practitioners' skills and interests resulted in 'considerable variation in [the] diagnosis' of asbestos and asbestos-induced lung cancer (1988, p5). These subjective accounts are then incorporated into legal processes as fact. Indeed, as Reinold Noyes argued in 1940:

> [T]he law is primarily a system of value judgement applied to human conduct. In spite of benevolent endeavours to expand its domain, it is quite clear, I think, to most scientists that science does not and cannot determine the judgement of values. Values depend on what (or which) we want. Science may aid us in getting what we want; it may clarify for us the actual alternatives between which we choose; it may show us the 'costs' of our choices; it may predict for us their consequences ... It follows that no application of scientific method can furnish the ethical code which a system of law expresses. (1940, p497)

For the laggers in particular, these legal categorizations are heavily influenced by political and economic considerations that seek to support powerful corporations and, in so doing, prevent the laggers from obtaining justice. It is perhaps not necessary to point out that the categorization of disease has massive implications for people diagnosed with ARDs. Consider the following comments, which come from the GMB laggers, and represent only a tiny sample of what I heard: 'They diagnose us with pleural plaque'; 'they are trying to diagnose it out of the system'; 'we do believe they are misdiagnosing us'; 'they say pleural plaque is not life-threatening, but what it develops into is'; 'Dr Dunne keeps fiddling us by telling us we have pleural plaque'; 'why are so many men diagnosed with pleural plaque and not asbestosis?'; and 'pleural plaque is a trick, they diagnose us with pleural plaque to pay as little as possible'.[5]

The following case of Erick Williams shows the complexity around medical diagnosis. As is evident in this case, it was sensations of pain which initially sent Erick Williams to his local doctor. These intermittent pains were, however, disregarded after the categorization of his illness as benign. The case also illustrates some of the grounds for laggers' suspicions of these medical and legal definitions that they encounter. In particular, the rapidity with which Erick's health declined and his subsequent death from mesothelioma, coupled with the way the legal system follows these medical categories, reinforces the laggers' belief that 'the system' is operating to hamper, rather than assist, their claims.

Case study: Erick Williams

In May 1999, Erick Williams visited his doctor and explained that he had developed pains in his chest. Because he was a lagger who had been extensively exposed to asbestos, he was sent for a chest X-ray. This revealed some 'pleural shadowing'. Eight months later, and after several more visits to his doctor, Erick was still complaining of intermittent chest pains. He underwent a second chest X-ray on 11 January 2000, which showed that there had been no change in his condition. As the pain continued, Erick underwent tests for ischemia[6] and had further X-rays and CT scans.[7] On 31 March 2000, a CT scan conducted by Dr Waterstone confirmed the presence of a 'left pleural node' which was judged 'most likely to be a pleural plaque'. A bronchoscopy[8] performed in April 2000 confirmed that there was no underlying mesothelioma. A repeat CT scan suggested that there had been no increase in the pleural node, which was, at the time, considered to be benign. These results were confirmed in June 2001 when Erick had another CT scan at Romford Chest Clinic.

In the meantime, Erick began legal proceedings. He was advised by his solicitors that 'straightforward claims' could be settled within a year or 18 months, whereas 'large or difficult claims' could take up to 2 or 3 years. Erick's solicitors requested that he undergo a medical examination by Dr R. M. Dunne, a leading UK physician and expert on lung cancer. This examination took place on 28 January 2002. Dr Dunne found no evidence of abnormalities in Erick's abdomen, no evidence of clubbing[9] and normal functioning of his pulse, breathing, heart and blood pressure. He also noted the presence of 'small uncalcified pleural plaques' but found no evidence of either pleural thickening or asbestosis. Accompanying lung function tests, not performed by Dr Dunne, also indicated normal lung function. Dr Dunne therefore concluded that Erick Williams was suffering from a 'minor degree of pleural plaque formation characteristic of asbestos exposure' that did not cause any disablement and did not affect his ability to work as a lagger. At the time, Dr Dunne estimated that Erick had a 3 per cent chance of developing lung cancer, a 6 per cent chance of developing mesothelioma and that his life expectancy had been reduced by two years. (R. M. Dunne, 31 January 2002, Medical Report for the Court, Ref W8979)

Dr Dunne argued that Erick's ability to work as a lagger had not been affected by the presence of pleural plaques. On the basis of Dr Dunne's report, Erick Williams took Falcon Insulations – his first employer in the lagging industry – and four other companies to court. In October 2002, the five defendants raised their initial offer of £14,442.15 to £15,000. Erick rejected this offer and, as a result, had to be re-examined by Dr Dunne. This examination took place in December 2002 and concluded that Erick's medical condition was 'unchanged'. In January 2003, Erick accepted a revised offer of £17,500 as a full and final settlement. This was an offer which was considered appropriate because it matched 'Counsel's[10] full valuation' of his case.

In January 2002, Dr Dunne had estimated that, instead of living to 80 or 81, Erick's life expectancy was reduced by two years. He confirmed this estimation in December 2002. This estimation was based on his medical examination of Erick, on the lung function tests and on statistical calculations using the projected life expectation tables. By all accounts, Erick Williams appeared a generally healthy man: he had stopped smoking 25 years previously and he had not had any serious illnesses, although he was slightly overweight for his height. The only problem was his ongoing complaints of chest pain. According to Dr Dunne's report, there was no reason why Erick should not continue working until retirement age. But things did not work out as predicted for Erick Williams. By April 2006 – less than four years later – he died of pneumonia and peritoneal mesothelioma.

The medical explanation of this case is clear cut: Erick Williams did not have mesothelioma at the times when he was examined. Because the average life expectancy, after diagnosis of mesothelioma, is only one year, it is entirely plausible that someone showing no symptoms whatsoever could be dead within two years. Nonetheless, this is not how the laggers interpreted this case and there are observations that explain their unease. In keeping with western medicine, Dr Dunne's report is 'single-case centred' and is based, not on laggers' cumulative experience, but on Erick Williams' personal experience (compare with Pfifferling, cited in Helman, 1984). As Stephens (2002) has pointed out, there is a big difference between statistical probability and predictability – on which scientific and legal decisions are based – and personal experience. In addition, everything about the medical process Erick Williams underwent emphasizes a detached, neutral approach. Pfifferling suggests that this might be termed a 'physician-centred' approach in that it is the doctor, rather than the patient, who determines what the problem is and its medical (and legal) severity (cited in Helman, 1984). Good demonstrates how people become objectified as projects and patients within the medical system, a process which 'justifies the systematic discounting of the patient's narrative' (1994, p78). Although Dunne makes reference to Erick's concerns about his ability to continue working, and recognizes the associated anxiety, these are largely irrelevant in the legal process. Indeed in the legal system, the patient as a formative agent has 'disappeared'.[11] Dunne's focus on physical/bodily health replicates the mind/body dualism so prominent in western medicine and is directed towards a legal audience. The nature of the expert witness in court cases reinforces the clinical professional knowledge of the specialist at the expense of the patient's more personal, experiential view *even when* the specialist is acting on the patient's behalf.

The laggers do not necessarily disagree with the medical categorization of ARDs, and they draw on this categorization in their quest for financial compensation. But the manner in which the expert witness is used in court cases reinforces the clinical professional knowledge of the specialist at the expense of the patients' more personal, experiential views. In the laggers' view, legal and medical authority thus manipulates the categories to deny that their personal experience of symptoms is 'real'. Their experience thus reflects western society's 'erosion in both the confidence placed in expert opinion, and the confidence placed by experts in their own assessments' (McInnes, 2005, p13; Durodié, n.d.). Linked to this erosion in confidence has been the declining status of knowledge, which is increasingly viewed as 'biased' or 'unattainable' (Palladino, 2002). As will become evident in the following section, this declining confidence in medical experts, strongly associated with the feeling of bias, is particularly evident in the laggers' collective experience of ARDs. Doctors are believed to conceal men's positive diagnosis in order to protect companies, to downgrade men's conditions to pleural plaques rather than admitting the full extent of their disease and to substitute older, healthy X-rays for recent, affected X-rays in order to avoid litigation and compensation. Their distrust is thus not of 'science' per se, but of the medical profession and the manner in which they perceive doctors to be colluding with the asbestos companies.

The laggers' view

In supporting the value of medical science, but questioning doctors' motives and the ways in which medical expertise is represented in court, the laggers maintain a very ambivalent relationship with the medical profession. At the heart of this ambivalence is the issue of validating what a 'real' disease is and locating their bodily experiences within these medical categories. In legal processes, ARDs and their resultant aetiologies are supported by psychiatric assessments. As discussed below, these assessments frame patient experience in terms of understandings of 'somatization', which reinforce the mind:body dualism inherent in medical science (Good, 1994; Kleinman, 1981). However, recent medical anthropology has contrasted the manner in which a patient's body 'feels' and 'demonstrates' distress with medicine's dualist understanding of the body. While biomedical understandings of pain overlook the psychological and social worlds of patients, Kleinman (1995) has suggested that bodily experiences of pain and 'social suffering' are integrally related to economic and other resources. Furthermore, both medical and social problems are mediated through the embodiment of distress. The laggers' account of the medical and legal processes thus disagrees with the underlying logic of disease categorization, with medical expertise regarding risk and statistical possibility, and offers a politically, economically and socially grounded understanding of the processes they undergo. These different perspectives on the psychological experience, engendered by both the legal process and by the fear of severe illness and death, further illustrate the complexity of categories.

As is evident from the quotes above and from Erick Williams' experience, laggers do not accept the categorization of pleural plaques as 'benign' or 'inert'. These terms are a mockery of their experience of pleural plaques: they 'diagnose us in the beginning with pleural plaque but it becomes asbestosis or mesothelioma either in the hospital or in the following weeks'. The laggers argue that the presence of pleural plaques on the lungs is an indication of their extensive exposure to asbestos. This is in keeping with medical definitions, but the laggers insist – contrary to medical opinion – that the pleural plaques point to the presence of other, worse diseases, as evidenced by the fact that 3 of the 12 men diagnosed in 1998 have subsequently died. The case of Erick Williams detailed above is a classic example. Furthermore, it is the very presence of pleural plaques which create stress and unease among the men. The chairman of the GMB explains: 'The mental stress caused by pleural plaque is very severe. About 80 per cent of the men diagnosed with pleural plaque die of asbestosis ... It's about the mental stress – they think they are on the way to mesothelioma.' Pleural plaques are thus directly linked to mesothelioma that, as a form of cancer, is highly stigmatized and widely associated with death (see, for example, Balshem, 1991). ARDs and cancer cannot be conceptualized in terms of gradation of illness nor as a short-term interlude in an otherwise healthy life. Not only is cancer frequently a metaphor of bad, unwanted experiences, it has been seen as 'naturally loathsome' and as an 'invisible contaminant' that invades the body (Bourke, 2005; Erikson, 1990).

The laggers – especially Ben Smythes, who was a close friend of Erick Williams – were not prepared to accept the medical explanation. Ben insisted on receiving the coroner's autopsy report, dated 24 April 2006, in which a number of observations were made. The coroner questions why Dr Waterstone's examination in June 2001, during which he confirmed the presence of a pleural nodule and argued that underlying mesothelioma could not be ruled out, was discounted. He also recommended a repeat CT scan in 12 months' time. It is not clear whether this repeat scan was ever carried out. The coroner argued that more attention should have been paid to Dr Waterstone's comments. Instead considerable weight was placed on the absence of pleural plaques and the lack of clubbing in Erick's hands and this led to a prognosis that ruled out mesothelioma.

This prognosis indicated that Erick's chances of developing mesothelioma were remote and it is on this basis that his case was settled as a full and final payment. However, as Erick Williams did die of mesothelioma, this prognosis was clearly 'misplaced'. The coroner's report therefore argues that the 'body of medical opinion needs to be revised' in order that future assumptions and diagnoses might be more accurate.

Whereas the coroner indicates a problem within medical opinion that requires addressing (although his report gives no indication of how this might occur), the laggers see medical opinion as a deliberate act of obfuscation on the part of medical science. Their belief is reinforced by the difficulty they experienced gaining access to Erick's records and by the doctors' and specialists' reluctance to share their information. Erick Williams decided – on the basis of Dr Dunne's repeated medical examinations – to take a full and final settlement in January 2003. Ben felt, however, that the defendants' doctor had access to Erick Williams' X-rays and correctly judged Erick's imminent death from cancer. If Erick had known this, he would have been able to sue for something in the region of £300,000 as his full and final settlement. The coroner is said to have told Erick's wife that the mesothelioma had been present in Erick's body for three-and-a-half years. 'He [the coroner] couldn't believe he wasn't screaming with pain. There was 70 per cent in his lungs and the rest in his abdomen.'

Dr Rajiv Menon is a psychiatrist who has been working on stress among UK asbestos sufferers and who, like Dr Rudd, frequently gives evidence in court on asbestos litigation cases. He points out that distress is not related to the severity of the illness and that the worst cases in terms of mental health are the people with pleural plaques. He identifies depressive episodes, anxiety and obsessive compulsive disorders with diagnosis of pleural plaques. In his work, he has documented men suspecting their wives of infidelity, following and accusing their wives, turning violent, being plagued by mental images, force-feeding themselves to put on weight, refusing to confront their own images in mirrors and suffering from phobias regarding dust and cleanliness (Menon, 2006). Nonetheless, his interpretation of the disease, although sophisticated in terms of understanding people's concerns, still reproduces a mind:body dualism of western medicine. In the above-mentioned High Court of England and Wales case, he argued that 'So when people like Mr Grieves [one of the claimants] were diagnosed with asbestos disease both reality and folklore fuelled ... fears and the psychological distress evoked by the diagnosis often outweighed the actual material risk of serious complications' (*Grieves and others*, paragraph 24).[12] In Menon's view, other clinical manifestations of stress include *disproportionate* breathlessness, *unexplained* chest pain, anxiety fears and extreme anger, thus suggesting that he accepts the medical model as normative. He concludes that there is little doubt that there is a psychological condition linked to mental health and those who fared badly were not 'psychologically minded', were emotionally vulnerable – particularly men who felt that they had to put on a strong face and be brave – and were unable to ventilate their feelings. In keeping with this strong, masculine identity, Dr Menon describes his patients as seeing him under sufferance and as not being willing to accept a psychological explanation for their physical experience. In his view, they find it hard to distinguish their physical conditions from their depression.

Before expanding on the laggers' rejection of psychological explanations, it is necessary to examine their second objection to the medical and legal definitions of ARDs. This is a critique of the political economy and powerful corporate processes that shielded asbestos mining and corporations in the past and that continue to hinder laggers' attempts to litigate for ARDs. As discussed in Chapter 1, asbestos was a wonder product and various companies made massive fortunes out of it, simultaneously creating hundreds of thousands of jobs and exposing those employed to asbestos (McCulloch, 2002; McClintick, 2000). As Jasanoff (1995) and Ward (2002) have shown, many of these companies are now facing large litigation and compensation claims. Society at large has also benefited enormously from the use of asbestos; in terms of fire protection and a myriad of other uses for asbestos, including insulation, brake linings and seals (Gee and Greenberg, 2002). In addition, nation-states have benefited from the economic success of the industry, from the use of asbestos as a fireproofing material and from its application during wars. Despite the high profile of asbestos litigation claims and the potential financial crisis of insurers such as Lloyds of London (McClintick, 2000), asbestos profits have 'suffered little from the ill health and contamination costs of asbestos, which were "externalised" onto workers with disease, their families, the health service, insurance carriers and building owners' (Gee and Greenberg, 2002, p58; Kazan-Allen, 2006).

Thus, while industry, nations and society generally have benefited from the use of asbestos, those people who worked with it – mining it, installing it, using it in manufacture and, more recently, removing it – and those who lived in the vicinity of asbestos companies, such as Cape plc in Barking, can be seen to bear both the social and physical costs. From the perspective of the laggers, the legal establishment, medical science, trade unions and the government are all working to assist business and industry. Their experience – watching family members suffer, dealing with doctors and lawyers – has led to deep suspicion of the medical and legal establishment. Ben Smythes and his colleagues are convinced that something is not right and that, in the case of Erick Williams, there has been a cover-up: 'The diagnosis that took place in 2001 should have said he had mesothelioma'; 'the doctor said in January 2006, you're in the clear, don't come back for the rest of your life, but the rest of his life was only three months'; 'they [the doctors] will not admit their faults, they will not admit to nothing'.

Because Erick settled for a full and final payment, nothing more can be done in terms of his legal claim for ARDs. Searching for new avenues of mobilization, the laggers considered suing the health authorities for incorrect diagnosis. They have approached the Law Society, the GMB and other knowledgeable people for guidance. They are determined to use Erick's experience to stop this happening to others: 'It shows what goes on, because this is going to come up again with other people, he's probably not the only one.' His friend Ben Smythes has taken a more personal approach. He wants the doctors to show more respect and suggests: 'If I can give the doctors one up, they won't be so fucking snobbish with others.' He is determined to do what he can for his friend: 'He's lying out there and he knows there's one geezer who is not fucking giving up and I will do what I can.' This view resonates with those recorded by Burnham (1982) and Brown (1979) on how modern scientific medicine might support capitalist interests and undermine people's health in American contexts. Or, as Jasanoff has argued, 'Politics is never far from view when one is observing science in action around topics of immediate social concern' (1996, p410). The laggers' understanding of their experiences is one that, in contrast to the medical and biological view emphasized by doctors, points to the underlying political economy of health.

Laggers, masculinity and identity

The dismissal of psychological values as influencing the perceptions of ARDs and pleural plaques are in keeping with the interviews I conducted. One lagger who had seen Dr Menon commented, for example: 'I found it a waste of time, it's not me mind. As long as I can do a day's work and it doesn't get worse, I'm OK.' If, as Menon argues, the laggers are 'not psychologically minded', it is worth exploring what other sociological, political and economic processes are impacting on their lives and how their identity shapes, and is being shaped by, their legal and medical experiences of asbestos.[13] Following Gutmann (1997), the remainder of this chapter links the laggers' psychological experiences with questions of power and inequality. This involves an analysis of laggers as gendered subjects through an exploration of their masculinity.

Connell (1987) introduced the concept of hegemonic masculinity to refer to men in positions of power, and patterns of practice that perpetuated men's dominance over women. He argued that hegemonic masculinities do not 'correspond closely' to men's lives. Instead they are models which convey fantasies, desires and ideals, offer guidance on relationships with women and 'articulate loosely' with men's construction of masculinities and everyday living arrangements (Connell and Messerschmidt, 2005, p838). Despite a wealth of research on hegemonic masculinities in relation to gender research, little has been done in terms of understanding the relationship between masculinity, science and technology (Mellström, 2002). Hegemonic masculinities, for Mellström, involve a 'certain bonding between men and machines grounded in the embodiment of technology' (2002, p463; see also Faulkner, 2000; Mollona, 2005). Although modern technology and hegemonic masculinity have a historical connection with industrial capitalism and, as Wajcman has shown, powerful institutions and political/economic interests underlie the use and control of modern technology, not all men involved in technology are equally positioned (cited in Mellström, 2002; Faulkner, 2000; see also Connell, 1987). It is thus necessary to distinguish different types of masculinities, and their associated degrees of power. Engineers are often depicted as 'powerful symbols of the equation between masculinity and technology', yet Faulkner argues that engineers are situated in a structurally ambiguous position, at the interstices of capital, labour and the state where they strive for professional autonomy while having to conform to the demands of corporate employers.[14] She also points to the class distinction between professional graduates based in offices, maintaining a remote connection to the artefact through computer technology and who have higher status than those engineers who do manual, greasy and dirty work directly on artefacts (Faulkner, 2000, p95).

Lagging is a dangerous, dirty and dusty profession. But it provided these men, many of whom were poorly educated and unable to achieve a profession, with a way to get ahead economically. As Jimmy Croft explained, most of the laggers meeting in the Dagenham Working Men's Club were 'from the East End, used in dockyard and dock works. They were not very well educated, I left school and I couldn't hardly read or write.' Erick Williams provides another example. At the age of 16, he left school and started work. Two years later, aged 18, he entered the lagging trade and started mixing lagging plaster (that contained asbestos) by hand. Lagging offered a way of garnering success outside the formal educational system. Laggers are officially classified as thermal insulation engineers (discussed in more detail below), but they do not need high academic grades or a university education to achieve this. In contrast to other engineers, trained in terms of objectivist rationality, emotional detachment and abstract theoretical approaches to problem-solving (compare with Faulkner, 2000), laggers are introduced to the trade through an apprenticeship, which appropriately trains them through bodily knowledge that is constantly enacted and practised. During this time, the men are exposed to the skills of the trade, beginning with simple tasks, learning how to work with their hands, getting to know the tools and their application. Mellström suggests that, among mechanics in Penang, Malaysia, this phase of apprenticeship is a 'never-ending process of continuous incorporation of bodily and cognitive knowledge … [that] involves seeing, listening, muscular exertion, touching, calculating, and not least competently practising' (2002, p464; see also Mollona, 2005). In addition to the long apprenticeship, it took nine years of work before one could acquire, through a combination of skill and knowing other men in the industry, a union ticket. As Connell and Messerschmidt argue, based on Bourdieu's concept of habitus,[15] masculinities should thus be understood to be 'configurations of practice that are accomplished in social action and, therefore, can differ according to the gender relations in a particular setting' (2005, p836).

Because of the skills associated with lagging and its classification within the engineering sciences, there is some prestige associated with it. Between the 1930s and the 1980s, laggers were paid as skilled engineers. They did pipework and ductwork. 'It was an absolute trade, you needed the skills.' As engineers, these men established themselves with pride. One of the key comments made by laggers, when asking them about their fathers, was: 'He was a proud man.' He was proud because of his ability to support his wife and family through his work as a lagger, because he was bringing young men into the industry, protecting them and supporting them, and – for many – because of the physical strength and fitness maintained through lagging.[16] The responsibility of 'breadwinner' and the moral dependability towards young men in the community reflects the significance of the patriarchal family and community structure. This esteem is reinforced by the men's physical toughness and mechanical skill which, Wajcman argues, creates one of the prevailing forms of masculinity in relation to technology, in that it allows men to demonstrate their power over, and manipulation of, both nature and technology (cited in Mellström, 2002; see also Hacker, 1989).[17] The embodiment of masculinity through technology is, as Mellström argues, 'extremely important in the professions of mechanics and engineers – professions in which the body physically enacts change on artefacts. The enculturation of the body, in both occupations, is to be seen as a never-ending process of continuous incorporation of bodily and cognitive knowledge' (2002, p464). Being a lagger thus involves developing a bodily knowledge which is performed and enacted and that constitutes their sense of masculinity and pride.

This sense of pride associated with a trade is, however, rapidly vanishing. Two processes in particular are eroding the laggers' hold on their profession. The first is the opening up of the European Union (EU) and the increasing numbers of foreign workers willing to take on work at considerably lower rates. The laggers talk, for example, about qualified Polish and other EU workers who underquote the UK laggers. The GMB has protested in relation to this, questioning foreign workers' qualifications, demanding that they be employed at UK rates and trying to ensure that UK labour is employed. In addition, the manner in which laggers subcontract work – operating as small, independent entrepreneurs – has disadvantages that, although they affect both local and foreign workers, are magnified in the case of men who are not familiar with the work sites and who are often not informed that they are dealing with potentially dangerous products. These foreign workers work short contracts before returning home. If they subsequently become ill, it will be virtually impossible for them to trace the connections between their condition, the products they were exposed to and their UK work contracts. Even resident laggers, when trying to process asbestos claims, often find that they have not been formally recorded on a company's books. Gavin Naul, for instance, was subcontracted to Cape, which paid his wages directly but which claimed that it had no record of his employment. This has implications both in terms of tracing employers in legal cases and in terms of any debates around company restructuring.[18]

The second process threatening the laggers' prestige is one that centres around the very nature of lagging. The Thermal Insulation Contractors Association (TICA) deals with all the terms, conditions and working rules for laggers. It also covers all the terms and agreements of engineering construction. Recently, negotiations have focused on the idea that lagging should not be categorized as 'engineering'. Currently most large lagging contracts have a large metal workshop situated on site. About 50 or 60 laggers will be employed and they will use this workshop to convert large sheets of flat metal into U-bends, elbows, piping and so forth to cover the insulation placed on piping. They call this process the 'making up of metal'. TICA, and many large companies that employ laggers, are considering removing the workshops and having all the metal 'made up' offsite by specialist firms. This deskilling of the trade[19] and downgrading of the machinery available upsets all the laggers; they argue that their expertise involves more than simply installing metal: 'We don't want to be downgraded, but that's what they are doing. They want to take metal synopsis off us. I find it demeaning. If you keep using a Stanley knife and not a piece of machinery, you're a tosser … If you go to work and you don't make metal up, you're a tosser.'[20] As Jimmy Croft suggests in this comment, not using heavy-duty machinery is directly linked to experiences of masculinity: a man who masturbates is not a 'real man', he does not have a wife and family. In addition, and as argued above, laggers' machines can be seen as symbolic extensions of their bodies. As Mollona has argued in relation to steel workers in Sheffield, these are 'metaphorical appendages of their male sexuality, and markers of social status' (2005, p533).

Negotiations, linked to this deskilling, are in progress with TICA. These negotiations are aimed at defending benefits the laggers currently have, rather than working to expand or enhance their role. A wide range of benefits, such as transport expenses, tea breaks and so on are targeted for cutbacks. The result, as one lagger explained, 'is a contradiction because it was a trade, it was excellent money, it paid better than other jobs on a building site. We became used to money, it was good money, but we're still earning what we were earning 15 to 20 years ago.' Thus, with a global credit crunch, real incomes declining, jobs being given to low-waged foreign workers, deskilling and unpromising negotiations with TICA, the laggers face an uncertain economic future, even if they remain perfectly healthy. These social processes – which reduce earning power, negotiating potential and the significance and pride in thermal insulation as a trade, and which enhance their economic vulnerability – have consequences for how the men think about themselves, for their identity and masculinity. Given the slow erosion in their position, it is not surprising that laggers no longer encourage their sons into the trade.

The laggers I interviewed expressed a sense of masculine identity based on two interrelated sets of activity: one concerning the 'public' sphere of economic work and the other located within the 'private' space of the family. Despite the widespread tendency in literature to see identities as multiple, hybrid, fluid and fractured (Connell, 1987; Moore, 1993), their masculinity was solidly grounded in their experiences as Dagenham laggers and as family providers. Like their fathers, these men identify as workers, husbands and fathers; they are fundamentally concerned with their ability to bring in money and to support their families. Thus, as Timmy Fortune commented: 'Our father made us a family unit, this has continued to his sons.' Such sentiments are in keeping with Gilmore's (1990) suggestion that masculinity is shaped through the ability to have children, to protect one's dependants and to provide for one's family or what Collier (1998) refers to as the 'positive' behaviour of men.[21]

The idea of 'the family' is broader than that conceptualized in a nuclear family and incorporates an extended, largely patrilineal, descent group. This family ideology has grown out of living in the East End of London and is, ironically, linked to the role of large firms such as Cape, Kitchings and so forth, which were seen as being 'like a family'. Many of the people working in these companies were related. In addition, the companies acted as hubs of social activity, they had good sports facilities, arranged sports days, celebrated Christmas and provided parties and gifts for their employees and their families. Until the late 1960s and early 1970s, when some companies ceased to operate, and others became increasingly contractual in their relations with employees, the companies were the primary employers in Barking and Dagenham and they looked after their workers in a paternalistic manner.[22] While the large companies no longer maintain this paternal role towards their workers and workers' families, the ideology of a family remains strongly located in both the laggers' sense of place and in their work. Mollona (2005) has, however, recently argued that extended families are being shaped by recent political and economic transformations in the labour market and the manufacturing sector as well as by changes in economic and social policies[23] at both the local and national level in the UK. His research, based on informal steel labour in Sheffield, suggests that patriarchal kinship relations have penetrated the factory floors or spaces of production while, simultaneously, capitalist relations of production have infiltrated the family and community and are orientating these spaces towards productive processes. He argues that deindustrialization has 'fostered household pooling, inter-household production and trading arrangements, and networks of "relatedness"... that extend outside the boundaries of the nuclear family' (2005, p542).

In contrast to the steel workers described by Mollona, laggers do not rely on informal work to supplement their wages. As Paul Long said: 'I've never known a lagger to do anything else, I've never known a lagger to work behind a bar or subcontract or try to supplement his wages through other activities.' Nonetheless, most of the laggers are 'normal working-class men' who have to support their wives and, as often as not, adult children. Although the laggers interviewed did not live in large extended families, neither did they necessarily resemble nuclear families. Ben Smythes, for example, is divorced. He lives with his adult son, who is also divorced. Ben spends several nights a week with his girlfriend and Ben's son's daughter stays with them every second weekend. Their need to protect and provide for 'the family' – which may include adult sons and daughters resident with them or divorced wives, sisters and brothers living elsewhere – is one of the primary reasons why the men insist on suing for compensation as soon as they are diagnosed, even if it is pleural plaques that they are diagnosed with.

'The good news is that Pat's going to be OK, the bad news is that they've offered me £96,000,' said Dick Fortune when he phoned his brother, Timmy. These two seemingly disparate comments refer to the same thing – the payment of £96,000. This money will help Pat, Dick's wife, survive once he is too ill to work and after his death; however, the fact that he is being offered £96,000 is indicative of Dick's poor state of health and highly reduced life expectancy. As noted above, it is the *diagnosis* of an ARD, even of pleural plaques – rather than the physical effects of the disease – which destroys the men's ability to continue working as laggers. This occurs in several, interlinked ways: there is the sheer number of family members who contract – and die of – ARDs that, in turn, results in increased tension within families when some people survive and others do not; there is the inability to support the family economically while the futility and inevitability of it all acts to close down discussion and opportunities for support.

Timmy Fortune explained how, in addition to his father, three of his brothers had died of ARDs:

> *First to go was me brother John, aged 66. He had been under the hospital quite a lot. He was a bit of a hypochondriac. No one believed him half the time. He said: 'This is the last time you'll see me' ... Next was my brother Dick. When he was aged 40 they said: 'there are shadows on your lung, but because you are so young, we'll do a biopsy'. Three months*

later, they said they hadn't got enough and cut him right open … Ernie died a year ago, aged 72. Ernie's daughter and son-in-law had taken Ernie and his wife to Turkey for a holiday. Within two days, Ernie fell ill and the local doctor in the hospital asked 'why are you travelling with mesothelioma?' Back home, he went straight into hospital and didn't come out alive.

The death of individual men threatens the broader family structure. Timmy Fortune keeps thinking about how his brothers died, about how Dick's daughter was on holiday when her father died. His decision not to interrupt her holiday and not to inform her of her father's death led to a family row which still permeates their relationship. Dick's grandchildren now ask Timmy: 'Why are you still alive and our granddad isn't?'

The inevitability and futility of their experiences operate to close down discussion. As is evident in Timmy Fortune's experience above, it is too overwhelming an experience to open up to discussion. There is no answer that Timmy can give his brother's grandchildren. There is no reason why he has lived, despite his exposure to asbestos, and his brothers have not. Indeed, as Timmy comments:

I've been too concerned looking after the immediate family to worry. It's like a graph laid out in front of you. You know exactly what's going to happen and they know what's going to happen. You have to put things in order, you can't ask 'Why? Why? Why?' What will be will be. That's the way I cope with it.

It is as if the very act of acknowledging the possibility of a disease may bring it on. This refusal to discuss the implications of diagnosis or potential diagnosis is part of remaining brave, of fighting 'it' both mentally and physically. For this reason, laggers avoid medical examinations. Indeed, checking for cancer is 'looking for trouble' and tempting fate (Balshem, 1991). The men know that their chances of getting some form of ARDs are high. As Paul Long put it, when we discussed whether my research should focus on men with ARDs or not: 'I'm sitting here waiting to be diagnosed.' This commitment to fighting the disease – before diagnosis – through refusing to acknowledge future possibilities, also extends to their families and home life. Edward Freeman recounted how most of the men working on a power station suspected that their wives and dependants might be worried, but that it was never expressed verbally.[24] His own wife knew of the dangers but said that he should make his own mind up and did not discuss his work with him. Even once men have been diagnosed with pleural plaques or other ARDs, they tend not to discuss it. As Paul Long said:

'My daughter doesn't know anything about it, my wife puts it to the back of her mind.' Timmy Fortune, whose loss of three brothers is described above, said that when he watched his sister: 'I can see her thinking "what's happening to all my brothers?" but we've never spoken about it.' He continued: 'Our family can't talk about it ourselves. They understand it's inevitable, we know its coming, so we only talk about it when it's inevitable. Then you've got to talk about it.'

Given the combination of pride, skill, physical fitness and economic prowess, a diagnosis of pleural plaques signifies a crisis in the laggers' ability to maintain their lifestyle and masculine identity (compare with Moore, 1993; Hearn and Morgan, 1990). A diagnosis of any ARD – no matter how benign – affects their earning capacity, their social relationships with other men, their roles as husbands and fathers and their masculinity. To begin with, their earning power: bringing in enough to live on and providing for a family requires more than simply the purchase of food. It is about a lifestyle. The laggers have become caught up in the housing boom and in the materialistic consumption of the 21st century. Despite the fact that the Barking and Dagenham housing prices are the lowest in London, many of these men battle to meet their mortgage and other daily expenditures. They desire, as one man put it, 'a decent lifestyle' that involves not being in debt and 'some money in me pocket'. In addition, the contractual nature of their business means that they do not benefit from the usual social protection mechanisms available. They have no sick benefits, no injury compensation, no means of surviving if they are not earning. Thus, if they are not working, they are not earning. Men who had been the mainstay of the family, proud men who had assisted everyone financially, found that they could no longer afford their own cigarettes once they stopped working.

The laggers meet weekly at Hamilton Hall pub in Liverpool Street, London, where anyone working in the area might drop in for a pint or two, and fortnightly at the Dagenham Working Men's Club. These meetings can be seen as 'truly gendered spaces' in which 'homosocial masculine practices continuously exclude women and perpetuate highly gendered social spheres, in which men form communities' (Mellström, 2002, p475). These are spaces for male bonding (Tiger, 1984/1969).[25] The GMB meetings are prefaced and concluded with drinks at the bar, general discussion and camaraderie. The meetings are where men maintain their relationships with other laggers, find out what jobs are coming up, who is working with who, who has not been paid and so forth. In essence, the entire conversation is related – in one way or another – to lagging.[26] Each week they discuss ARDs: who has been diagnosed, who has not, what the current status of each lagger's claim is, who has died and whose funerals have been (or have to be) attended. Although the mood is generally jovial and the laggers are full of ironic humour when discussing their problems, there is also a strong recognition that sooner or later they too will be affected. Laggers who have been diagnosed find themselves being constantly watched by the other laggers for symptoms. Every time they come to the meetings, they are reminded of their own situation and they hear about more people dying. Steve Harris, for example, was undergoing medical examinations for unexplained pain and for a tumour in his abdomen, while the men gathered to discuss the case of Erick Williams. Although he has not formally been diagnosed with any ARD, the 'man' who did his X-rays asked if he had been exposed to asbestos. When Steve replied that he was a lagger, the man rolled his eyes. Steve then asked if there were signs of asbestos, to which the man replied 'yes'. Although Steve tries not to dwell on this experience, at the time he compared his tumour to the mesothelioma 'mass' in Erick's stomach. 'I was coming to the meetings, but I hadn't told any of the men.'

As suggested in Steve's comment above, the laggers seek to manage the process as far as possible. The disclosure of an ARD is carefully handled, with each individual deciding how to disclose the information of a positive diagnosis and to whom. This is because the diagnosis of an ARD, even of pleural plaques, may signal the inability to continue working and many men stop attending GMB meetings once diagnosed. This means that they lose touch with their lifelong friends and 'family', separating themselves from their support structure. Disclosure of a diagnosis is thus closely related to the destruction of the social person, through a process of self-rupturing that pre-empts the disruption of the body. Moreover, there is no other support structure to which they can turn. Faulkner suggests that engineers' intimate identification with work and its associated technology provides engineers with a 'separate reality' as evidenced in the meetings. Furthermore, it prevents them from asserting more positive identities when among non-engineers. 'Indeed, we may view engineering as a fraternity built around this common identity with, and pleasure and pride in, technology' (2000, p107). Men who have been diagnosed demonstrate this extreme dislocation bodily: they are seen to change in personality, withdrawing into themselves: 'They all get the same attitude, they jump at anything, they look out the window, they don't talk, they go into themselves.' As Menon (2006) points out, since the 1970s these men have been aware of the dangers of asbestos and have been waiting for the diagnosis. Once they get the disease, it is a catastrophe for them.

Ultimately, the laggers find themselves in a complex situation. On the one hand, the men avoid seeking medical help and confronting diagnosis for all the above-discussed reasons. A positive diagnosis is not merely about acknowledging the disease, it also initiates withdrawal and social and professional 'death'. On the other hand, and as will be demonstrated in the following section, the laggers take on the legal fight as a fight against authority. Their legal struggles have a communal component and become a further expression of their masculinity.

Masculinity: Family and legality

Until January 2006, following the Rothwell decision in the Court of Appeal, laggers would initiate a claim as soon as they were diagnosed with an ARD. Their approach to this legal process has been remarkably uniform. Although the legal framework compensates more to people who have more severe forms of ARD and although laggers are convinced that pleural plaque is an indication that they will succumb to asbestosis, lung cancer or mesothelioma in the future, they all agree on a full and final settlement. As illustrated in the case of Erick Williams, who was paid £17,500 in a full and final settlement for pleural plaques and then died of mesothelioma less than four years later, this appears to be strangely illogical. It certainly runs counter to the legal position that people will seek to make a business out of litigation and the idea of a compensation culture which aims to maximize financial terms. The case of Ben Smythes, the only lagger I came across who took a provisional settlement, shows how one might benefit from such an agreement. In 1987, he was diagnosed with pleural plaques. When he went to court, Dr Dunne estimated the degree of damage to his lungs to be 20 per cent while the opposition doctors said 10 per cent. They settled on 15 per cent and Ben received £40,000 as a preliminary settlement for pleural plaques in 1987. As he puts it: 'I lived to fight another day.' There has been a slight deterioration in Ben's condition since then but no significant changes in terms of the medically defined categories of ARDs. Should this happen, Ben can go back to court and claim further compensation. From an outsider perspective, this seems the most lucrative and logical way of progressing, but further investigation into Ben's case shows that this is not what he intended: he initially went to court hoping for a full and final settlement, but when he discovered that the defendants had included a second doctor's opinion (based on his medical records rather than a physical examination), he became worried. Not only did this doctor indicate that it was Ben's lifestyle that was problematic by including a diagnosis of obesity in his report, but 'I was going to get slaughtered, it was one against two'. Ben asked for an adjournment– in order that he might have time to find another doctor to support his evidence – but this was refused. He then requested to change his claim from full and final to preliminary, aiming still to find additional medical support for his claim of 20 per cent. It was on this basis that, once the case resumed, Ben was stuck with the preliminary claim. Today, aged 61 and reflecting back on the case, he says: 'I don't know why I'm not dead, I shouldn't be alive now.'

It is for precisely this reason that the laggers chose a full and final settlement. Once diagnosed – even if with only pleural plaques – they have no way of knowing how much longer they are going to survive and they believe that the next phase for them is 'in a hospital bed'. So the men, who think that they are going to die anyway, aim to get the legal process over with and to collect a compensation payment as soon as possible. Indeed, a legal claim can take anything from two to ten years to complete, leaving the men in limbo throughout this time. Their intention is to invest this money as a source of protection for the family, and simultaneously to find a new, easier way of earning an income. This is because, despite medical opinion, many men experience breathlessness from pleural plaques and find it hard to continue the gruelling workload associated with lagging. They hope to continue to be able to bring in some money, albeit less than before: 'If you get a lump sum you can diversify, get a different job, take a different course in life' but 'you can't go back, you get your bit of dosh and go and do whatever'. However, although Ben (described above) received £40,000 for pleural plaques in 1987, compensation payments have been going down steadily since then and the average payment in 2005 for a case of pleural plaques was about £6000–7000 (O'Neill, 2005). Although this is not enough to invest – and to guarantee – a family's economic security, the money is still used to 'put your house in order'. Almost all the men interviewed used the money to pay off their mortgage, to buy new furniture or for a holiday.

Conclusion

Ironically, the laggers' approach to ARDs is one that enhances their masculinity. They 'fight' the disease through refusing to admit its presence, through refusing to discuss it with their wives and other family members, through engaging in legal court cases and through fighting their deceased friends' cases. In this context, the act of suing is also an expression of masculinity. This fight is one in which, although couched in the language of financial compensation, is fundamentally about issues of recognition. It is about the acknowledgement of the laggers' experience, about the recognition that pleural plaques are 'real' and have immediate ramifications. The fight against the medical and legal systems can also be viewed as a metaphor for a certain way of life. It is about the survival of people who have old industrial diseases, about men whose lives are at an end, who are part of the old white working class.

There is a postscript to this story: the laggers' concern with Erick Williams' experience led them to request that their solicitor obtain further clarity and 'missing' medical records from Dr Dunne. Dr Dunne responded to a series of specific questions, provided the 'missing' medical reports and explained his reasoning. Finally, he said:

> *I can assure the authors … that there is absolutely no basis whatsoever for the suspicion that evidence of mesothelioma should, or even could, have*

been detected in January 2002 or in December 2002, the two occasions when I saw Mr Williams. I estimated a 6% risk that Mr Williams would develop mesothelioma and that can emerge at any time.[27]

Both Dr Dunne and the laggers' solicitor, Martin Longsmith, sought to reassure the men that pleural plaques were not precursors of other, more debilitating ARDs. Martin Longsmith wrote, in his covering letter to the president of the Dagenham branch of the GMB, 'It further follows that members of your branch who currently have pleural plaques need not fear that they also have mesothelioma although, sadly, a minority are bound to contract it in the future' (8 November 2006). His response adheres to medical and legal processes which seek to turn uncertainty into ideas about risk and certainty. For this reason, the 6 per cent has come to hold a particular significance. As is evident in the above discussion, the laggers are also working with ideas of certainty and argue that 80 per cent of the men diagnosed with pleural plaques end up with mesothelioma. It is perhaps for this reason that, when reading out Dunne's reassurance to his members, the president misquoted Martin Longsmith and said: 'It further follows that members of your branch who currently have pleural plaques need not fear that they also have mesothelioma although, sadly, *a majority* are bound to contract it in the future.'

What the president did not read out, was Dr Dunne's conclusion:

In my view, the moral to be drawn from this sad case is not that mesothelioma should, or even could, have been detected in 2002 but that individuals in Mr Williams' position should give very serious consideration to provisional damages rather than final damages because it is simply not possible to know if and when mesothelioma may emerge and if persons who have had heavy exposure to asbestos take final damages for benign conditions it is inevitable that a few of them will develop mesothelioma not long after their cases have been settled.[28]

Biomedicine, law and engineering all reproduce a mind:body dualism that favours rational, unemotional, individual and positivist ways of problem-solving that are, as Good has pointed out, based on the objectification of the body (1994). These dualisms, coupled with the biologically dominated perspective of medical science, are perpetuated in the legal process of obtaining compensation for ARDs. Simultaneously, these processes negate the experience of laggers, who are physically involved in their trade, who experience their masculinity through the use of machinery and through the processes of male bonding and who therefore experience pleural plaques as a form of bodily incapacitation despite medical expertise and their 'benign' nature. The diagnosis of pleural plaques represents a crisis in their masculine identity. Dr Dunne's attempt to resolve the controversy around Erick Williams' death through reference to individual responsibility and the way the laggers make use of the legal system is not surprising. It is part of the way medical practice prioritizes 'rational', individualistic behaviour over social constraints and intersubjectivity (Good, 1994, p23). However, as this chapter demonstrates, such an approach fails to recognize the manner in which laggers construct their masculinity around medical/legal frameworks and the socio-economic and political environments in which they experience the disease and enact their masculinity.

The current legal position, based on 'scientific medical evidence' that pleural plaques are benign and that sufferers cannot sue for compensation, removes the opportunity for laggers to claim and invest money – while still alive – to provide for their families and, in so doing, to fulfil their role as men. This, in turn, increases – rather than reduces – their stress as they struggle to maintain their identity and role as men, laggers and providers. As is discussed later in this book, the laggers were, at the time of writing, protesting for a change in the law, mobilizing alongside trade unions and international ban asbestos organizations for Members of Parliament to recognize their condition. The following chapter explores a rather different context in South Africa, focusing on the rural towns where asbestos mining has resulted in massive environmental pollution and in which ARDs are widespread. In keeping with the main themes explored in this chapter, the role of medical science and the notion of risk is contrasted with sufferers' own understandings of diseases and hazards. In contrast to the dominance of medical and legal expertise in the UK debates, working together with scientific experts has facilitated a process in which community participation has – to some extent – reshaped scientific knowledge for both scientists and town residents. Through engagement in asbestos issues, sufferers and other town residents in South Africa have sought to shift their identity from victims to active participants in policy processes who, like the laggers, seek their own ways of guarding against risk.

Notes

1   Warner (1995) argues that the term 'science' refers to cumulative knowledge about nature and to claims made in the name of science. This suggestion, that science is both discovery and an authority that extends beyond technical knowledge, is particularly pertinent when medical and legal domains interact and when science is vested with the power of classification (Kohlstedt and Longino, 1997).

2   Steele and Wikeley (1997) point out that the 1931 regulations had little effect. They were applicable only to people working in specific scheduled activities within the factory buildings while excluding other factory workers and residents living in the

vicinity of the factory. In addition, companies did not comply with these regulations and the government did not enforce them. Working specifically on Turner & Newall, Tweedale (2000) argues that not only did the company not comply with health and safety regulations, it also neglected to warn employees about the possible dangers, used a variety of methods to hinder legal processes and suppressed medical evidence.

3   www.nhsdirect.nhs.uk/articles/article.aspx?articleId=35&sectionId=10122 accessed 13 June 2007.

4   http://ibasecretariat.org/lka_ards.php accessed 13 June 2007.

5   De Vuyst and Gevenois point out that improved medical technology, particularly high-resolution computed tomography, has enabled doctors to detect subclinical cases of pleural lesions and has, as a result, led to fewer diagnoses of asbestosis (2002, p145).

6   Restrictions in his blood vessels and blood supply.

7   Known as computed tomography, a CAT or CT scan uses mathematical and computer technology to transfer information derived from two-dimensional X-rays into a three-dimensional image of the internal organs.

8   This technique of inserting a small scope into the body (often through natural body openings) allows for the examination of the lower airways and for the taking of samples, while ensuring minimum discomfort.

9   Clubbing involves a thickening of the fingers and a softening and rounding of the fingernails. It is associated with a range of heart and lung diseases.

10 Counsel refers to the barrister who gave expert advice to Field Fisher Waterhouse on this case.

11 In his historical examination of colon cancer, Palladino takes issue with this perspective and argues that reforms within the NHS position patients as 'informed consumers'. This transforms them from passive objects within the clinical gaze to 'constitutive figures of a new "discourse"' (2002, p139).

12 In fact, Mr Grieves' condition improved without medication, much to Dr Menon's surprise, and this is said to be in part because he had reflected further on Dr Dunne's diagnosis.

13 Consideration of laggers' exposure to asbestos, particularly in the US (for example, Selikoff et al, 1964, 1968; Selikoff and Hammond, 1975; Selikoff, 1977), is common in the academic literature, but very little attention is given to the sociological impacts of disease or laggers' identity more generally.

14 Engineers' collective association with technology is thus seen as a means to present themselves in a neutral light.

15 Bourdieu examined how people embodied social discourses and how they, through practice, translate discourses into potential or into limiting repertoires.

16 Many of the laggers followed their fathers into the trade and all of them did an apprenticeship, usually lasting about five years. During this time, strong bonds were formed between the apprentices and the laggers. A lagger would often look out for his apprentices: protecting them from bullies, from exploitative or dangerous jobs and helping them find secure work once their learning was complete.

17 Wajcman sees the other dominant form of masculinity to be established through 'professional calculative rationality' by men who are trained as technical specialists (cited in Mellström, 2002, p462).

18 Companies facing litigation around asbestos issues are well-known for restructuring as a means of securing their business interests and removing the risk of going bankrupt through litigation. This involves, in part, putting aside a trust fund ostensibly to ensure that future claims will be accommodated. However, as illustrated in the debates around Cape's restructuring in 2006, this is a way of limiting damage with no formalized means of topping up the trust fund once depleted. This process therefore allows companies to place a ceiling on their asbestos compensation expenditure with complete disregard for the numbers of people that they may have exposed to asbestos.

19 Faulkner also points to a general demise of the 'golden age of engineering' as engineers are no longer highly respected as pioneers of technology, but are more likely to be seen as nerds (2000, p104).

20 Often linked to the term 'wanker' and masturbation, 'tosser' is UK slang for an incompetent, disliked or stupid person. It can, however, also have positive, affectionate connotations.

21 Collier is responding to a tendency in the literature that views hegemonic masculinity as predominantly negative, associating men with dispassionate, autonomous neglect of the family and antagonistic behaviour. Connell and Messerschmidt (2005) point out, however, that this tendency is associated with personality trait theory and that other interpretations of hegemonic masculinity acknowledge that men's behaviour can work to women's advantage.

22 Wikeley, examining the case of Turner & Newall, argues that the employers' commitment to this moral and paternalist style of management also worked to eliminate any suggestion that they might be poisoning people (1997, p268; see Bartrip, 2001 for a sympathetic reading of Turner & Newall's activities).

23 These include housing benefits that do not incorporate necessities such as gas, electricity, water and building repairs, and the reduction in welfare benefits for former industrial workers. The limited financial support provided by these policies force families to provide financial and residential assistance (Mollona, 2005, p540).

24 Komarovsky (1967), researching masculinity in an American steel town, found a similar lack of communication between working men and their wives.

25 Although the branch has two women members and I was permitted to attend all these meetings.

26 Even when I tried to initiate broader discussions on the men's family relationships during these meetings, they quickly changed the subject back to either their own formative experiences as apprentices or to their current work problems or experiences.

27 B. Dunne, private letter to M. Longsmith, 1 November 2006, p2.

28 B. Dunne, private letter to M. Longsmith, 1 November 2006, p2.

Chapter 3

Evaluating Science and Risk: Living with and Dying from Asbestos in South Africa

Introduction

South African compensation for asbestos-related diseases (ARDs) takes a different format to that experienced in the UK and described in the previous chapter. Within South African medical and legal circles, there is widespread recognition of pleural plaques as a disease that is liable for compensation.[1] Nonetheless, scientific understandings of ARDs, including pleural plaque, and of risk still do not conform to people's own experiences and understandings of asbestos hazards. South African residents' mobilization and awareness of asbestos issues have worked, not only to secure compensation and challenge legal structures, but also to improve community participation and relationships between community structures and scientific experts. In so doing, asbestos sufferers and town residents have – to some extent – been able to shift their identity from that of victim to active participant in policy processes and to work with scientists in order to protect themselves against risk. This chapter explores this process of participation, focusing particularly on how mobilization can reshape scientific notions of risk towards community concerns. Two contrasting interpretations of asbestos risk and harm in the Northern Cape, South Africa, are examined, namely, a medical, scientific discourse about the probability of risk and an emic viewpoint regarding bodies, relationships and emotions. Although they appear as separate, unrelated interpretations, the cultural embeddedness of 'the public' and the fact that society is 'mutually implicated in science' (Michael, 2002, p374) are taken as a starting point for the analysis. This mutual implication is evident in public participation in policy forums and in local communities' engagement with science in hospitals, at funerals and when removing exposed asbestos.

Research into science and technology has examined how concepts of risk and public understandings of science might be useful in non-western contexts (Fairhead and Leach, 2003). Pioneering work on how scientific knowledge is embedded in – and interacts with – a cultural context was done by Douglas and

Wildavsky (1982) who argued that understanding risk requires an awareness of the socio-cultural context alongside knowledge of subjective and personal factors. Wynne's interpretationist approach also focuses on context and provides a means to re-examine the more traditionalist, positivist survey approach to public understandings of science (1995). His work draws attention to the interaction between 'formal' and 'informal' forms of expertise, showing how on the Cumbrian fells, sheep farmers interact with physicists. He draws attention to the relational dimension of risks, asking 'how far might lay people be involved in shaping scientific knowledge, and thus in providing the basis for alternative forms of public knowledge that reflect and sustain different dominant conceptions of the human, and of the social purposes of public knowledge?' (1996, p61). More recently Furedi has reinforced the centrality of context with his argument that '[f]ears about the future are linked to anxieties about problems today. And if the future is feared, then reaction to risk is more likely to emphasize the probability of adverse outcomes' (1997/2002, p18).

In contrast to the science and technology debate, much of the literature on participation and citizenship has focused on developing contexts and on enhancing democracy and governance (Gaventa, 2002; Cornwall, 2004). As Leach et al argue, there is a need to 'explore the cross-context "translateability" of theories and debates and the possibilities of cross-learning' (2002, p41). In exploring contrasting conceptualizations of harm, this chapter argues that scientific understandings of risk and a 'public understanding' of risk are not mutually exclusive. Indeed, it is specifically through opportunities for participation and collective action around scientific issues that both scientific and lay understandings of risk are broadened. Using asbestos exposure and the risk of contracting ARDs in a non-western context as the site of investigation, this chapter argues that evaluations of risk are not always negative, that risk is not always linked to probability and is not always forward-looking. It furthermore shows how, contrary to many studies of public understandings of science which keep 'science and nonscience distinct' (Michael, 2002, p359), notions of science, risk and bodily harm shaped during processes of participation and political decision-making are interwoven with everyday knowledges and experiences.

The history of asbestos mining

South African asbestos production – and subsequent environmental pollution in the form of widespread asbestos waste – was located in rural areas. Although there were asbestos mines in several different parts of South Africa, this research has focused on the Northern Cape, and in particular on the areas of Prieska, Koegas and Griquatown. Asbestos mining was first initiated during the era of British colonialism. In the Northern Cape, for example, Cape acquired a crocidolite mine in Koegas and a mill for crushing asbestos in the town centre of Prieska in 1883. The Cape-owned mines Koegas, Elandsfontein and Blackridge were among the largest Northern Cape asbestos mines (A. L. Hall, 1930). Initially, all mining was done manually without compressed air and jackhammers, which meant that a man could take a whole day to drill a hole 30cm deep. After blasting, broken rock was removed using picks and shovels and the asbestos separated by hand. Underground seams were also worked by hand, by candlelight. After the invention of the pneumatic drill, conditions changed with conveyor belts transporting the ore and with primary crushers breaking up the rocks (McCulloch, 2002). Open-cast mining, in which the earth was dislodged with dynamite, became the primary means of extracting asbestos-containing rock. Thereafter, removing and sorting the asbestos fibres was done by women, often sitting breastfeeding or watching their children playing nearby while cobbing[2] asbestos fibres. Asbestos workers and their families were thus in daily contact with the hazardous material while the asbestos industry, supported by senior personnel in the Department of Mines, argued that asbestos was not carcinogenic. This alliance between industry and government labelled growing international evidence of asbestos hazards as 'wrong and prejudiced publicity' and as 'propaganda' (McCulloch, 2005, p8).

Economic relationships between the UK and South Africa continued to facilitate the growth and expansion of South African mines producing asbestos for UK companies (McCulloch, 2002). By 1977, asbestos production peaked at 380,000 tons, making South Africa the third largest supplier in the world (Kazan-Allen, n.d.). South Africa thus witnessed a century of asbestos mining. During this time, asbestos mining companies extracted raw materials in South Africa – with scant regard for the safety of its workers – and processed manufactured goods in London's industrialized zones, showing similar disregard for the resident population, until high levels of ARDs forced the closure of UK factories in the 1960s (McCulloch, 2002; Meeran, 2003; Gravelsons et al, 2004) and the UK ban on blue asbestos – as produced in the Northern Cape mines – in 1985. Shortly after this, all the Northern Cape mines were closed, either because they were worked out or because it became impossible to continue to ignore the medical dangers of asbestos, as asbestos litigation in the US expanded (Jasanoff, 1995) and highlighted the links between blue asbestos and mesothelioma (McCulloch, 2002).

When the mines closed, many mine residents and employees moved to nearby towns such as Griquatown and Prieska. In 1985, these towns were declared 'dust control areas' by the apartheid National Party government because of the widespread distribution of asbestos waste. Although asbestos waste dumps should have been covered with soil ('Distrikstrekordboek', 1978), this did little to help protect people from exposure to waste asbestos. About 20 years later, Kazan-Allen reported that: '[l]evels of environmental pollution in some areas are colossal; huge asbestos dumps, including eighty-two in the Northern Cape alone, are scattered throughout the countryside. To date, the current democratic government has spent R44 million ($7.5 million) on rehabilitation of derelict mines. Estimates that R52 million is needed to complete the task seem optimistic' (Kazan-Allen, n.d.).

Prieska is an economically active rural town positioned on the banks of the Orange River. It has abundant irrigated land that enables the production of maize, wheat, lucerne, peanuts and fruit farming alongside high-value crops such as pistachio nuts, olives, figs and pecan nuts on privately owned commercial farms. Sheep farming and game ranching occur on unirrigated farms. Industrial development includes salt pans, leather tanning, meat production, a cotton mill, the manufacture of cattle food pellets and a semi-precious stone industry. Former asbestos mine workers and their families have, however, only very limited access to these economic opportunities as racial and class hierarchies established during the apartheid era continue to predominate. Their proletarianized status as former mine workers and lack of access to land or resources means that they experience high levels of unemployment and poverty, despite Prieska's economic strengths.[3] Griquatown has even higher levels of poverty and unemployment than Prieska, as the primary economic activities of sheep farming, game ranching and a small semi-precious stone industry employ very few workers. These town residents largely survive through state payouts for old-age pensioners and disabled people. So prevalent is this aspect of the town's survival that residents refer to Griquatown as the 'pension town'.

Companies such as Cape failed to invest in the towns where their mine employees moved to or to consider the impacts of asbestos mining on local people. Instead, they responded to evidence of asbestos dangers by developing a two-pronged approach: as described in Chapter 2, in the UK they ostensibly emphasized factory safety and new technologies of dust control while simultaneously, in South Africa, disguising employees' ill-health and downplaying the scientific evidence of ARDs (McCulloch, 2002). As Myers has shown, the standards used in South Africa to define 'safe levels' were not based on debilitating ARDs such as mesothelioma, but rather on benign pleural disorders; thus justifying extensive exposure to asbestos fibres (1981, p231). The collusion between the South African government and UK asbestos mining companies was further enforced during World War II when companies were granted reprieves from state legislation (Braun et al, 2003, p194).[4] In addition, asbestos companies described the extraction process as 'not strictly mining', thus evading the mining legislation (despite opposition from the Departments of Native Affairs and Health). This, coupled with financial support to and control over medical research, enabled mining companies to ignore the dangers of asbestos and to oversee an enormously profitable extraction process (Felix et al, 1994; McCulloch, 2002, 2005).

Thus, while industrialized countries experienced public contestation regarding the hazards and uncertainties associated with technological processes and scientific developments (Castleman, 2001), asbestos mining flourished in apartheid South Africa (McCulloch, 2005).[5] Today, South Africans have little doubt about the dangers of asbestos. In addition, asbestos mining no longer takes place and the democratic government has banned the use of asbestos, insisting that its disposal also be carefully monitored. The following section examines the social mobilization that created this awareness and achieved these gains.

Creating political awareness

In March 2003, a small community group won a UK court case against Cape plc that, in July 2003, transferred £10.5 million to the lawyers representing 7500 South African claimants. The community group, known as the Concerned People Against Asbestos, or the CPAA, was based in Prieska, a rural Northern Cape town, and was organized when, in 1979, the surrounding asbestos mines began to close. The CPAA initially campaigned for improved access to compensation for asbestos diseases because, for people living in rural areas, this was a bureaucratic nightmare. In addition to having to travel long distances for medical certification of their disease, they had to complete various detailed questionnaires and negotiate with the Medical Bureau for Occupational Diseases (MBOD) based more than 600km away. The lack of access to transport facilities, to telephones, to their own medical records (often held by doctors in neighbouring towns) and to money, coupled with very high levels of illiteracy, made it extremely difficult for former Northern Cape mine workers to negotiate with the MBOD and to secure compensation.

Because Prieska had been centrally located within the asbestos mining industry for more than a century, all its residents were affected by asbestos. Although the CPAA was made up of ordinary people, its leadership tended to be educated, exposed to alternative and radical ideas and employed in relatively well-paid positions (see Chapter 6 for more discussion of leadership). They were concerned about the unreclaimed asbestos sites and piles of asbestos waste, unemployment (after the mines closed) and, as the latency period lengthened, the increasing numbers of people diagnosed with ARDs. With time, and with the democratic transition, the CPAA campaign became an opportunity for community representatives to voice their opinions regarding community issues and to be politically engaged in decisions about Prieska (see below). As one

CPAA member explained: 'Because, for once, after the elections we had the right to speak. We weren't suppressed anymore.'

Prior to the Cape plc court case, the CPAA had secured some victories in relation to the MBOD and access to compensation (discussed in detail in Chapter 5). However, the MBOD's general failure to appreciate their concerns led them to tackle the rehabilitation of asbestos-polluted areas with various government departments and to create networks with trade unions and international, northern-based non-governmental organizations (NGOs) and activists. During these early stages of campaigning, it became increasingly obvious to the CPAA that the new South African government, although supportive of their cause, was financially unable to address the multifaceted asbestos problems. The CPAA's campaign began, therefore, to focus more on international possibilities and, with assistance from various government officials, contact was made with the international human rights lawyers who eventually took on the Cape plc case.[6] The CPAA, already sophisticated in its dealing with South African government officials, now developed new skills. CPAA activists improved their mobilization strategies through international networking and through the new perspectives that the lawyers brought with them. Working with the lawyers, they discussed different possibilities and learnt to use political support more creatively, to publicize medical evidence and to be strategic in their use of media: 'We had to think through,' explained one CPAA campaigner, 'how can the political be used? How can the doctors be used? How can the claimants be kept happy? How can the law be used? And all these things had to be juggled simultaneously.' The lawyers, working with the CPAA, ensured that the political spotlight was kept on asbestos issues. ARD sufferers were regularly featured in media articles, certain individuals were chosen to give testimony in the High Court of London during the Cape plc hearings and the CPAA continued to mobilize local support to address diverse aspects of the asbestos problem. This led, in turn, to a reconsideration of earlier understandings and a new awareness of how politics and established power relations had influenced the decisions made by asbestos mining companies and governments. The Cape plc case came, in Prieska, to be seen as something new, as a way of challenging mining companies and as a means of empowering local residents.

Thus, while engaging in the Cape plc court case, the CPAA sought greater participation in local policy processes. Throughout the Cape plc court case (which lasted more than four years), the lawyers funded a CPAA office in Prieska – equipped with a telephone, photostat machine, computer and with access to email – giving these activists a substantial physical and material presence within the town.

The CPAA also sought to estimate the extent of the disease in Prieska, and to reach its own conclusions about the degree of hazard posed by asbestos. In so doing, it drew both on scientists' medical assessments and on more personalized notions of harm. For example, in conjunction with the National Union of Mineworkers (NUM), it invited the Industrial Health Research Group, based at the University of Cape Town, to 'review the occupational histories and audit the chest X-rays and lung functions' of mine workers working on an asbestos mine near the town of Kuruman (IHRC, 1996). This not only produced medical evidence of the prevalence of ARDs, but also recorded former mine workers' deep anxiety (IHRC, 1996, p2). Throughout its engagement in policy processes as explored in detail in the following sections, the CPAA has sought to integrate scientific discourses with community understandings of health and risk.

Asbestos mining and rural participation

Richards has argued that participation can be understood in two forms: deliberative participation and participation as performance. Focusing on the latter, in which participation is seen to include action, ritual, belief and performance, Richards argues that the participatory event has the capacity to 'generate excitement relevant to the fixing of collective representations' and, in so doing, produce new forms of action, belief and commitment (2006, p3). In an examination of agricultural innovation as performance, Richards shows that 'Interaction between people with different kinds of knowledge and different areas of experience can be a necessary stimulus' (1989, p46). It is through the 'doing of' technology, through action, that material outcomes and social values are achieved. Drawing on this knowledge-as-action approach, this chapter demonstrates how different ways of conceptualizing harm and risk – in particular, local cultural understandings and scientific approaches – interweave and support each other at times, but can also present contradictory and divergent interpretations of danger. Through an examination of medical screening, environmental rehabilitation and community development, the following section demonstrates how scientific knowledge is interwoven with activities and therefore skill-based. It shows how concerted action by agencies and local communities can lead to a broadened understanding of risk which draws on both scientific understandings and community perspectives of risk.

The rural town of Prieska is perhaps most closely associated with asbestos issues today. This is because it has featured prominently in the media on asbestos pollution and as a site of social mobilization against asbestos (as highlighted in the Cape plc legal case described above). The close proximity of asbestos mines, dumps and the former milling of asbestos in the centre of the town has heightened the likelihood of people contracting ARDs.[7] Only after a decade of the CPAA's concerted activities to address the health, environmental and occupational spread of asbestos issues, and after the 1994 transition to a democratic government, was there a concerned and official attempt to address asbestos pollution as a provincial concern. The African National Congress-led government also sought to encourage local level community participation through the Municipal Structures Act, through Local Government Councils, through the 'constitutional requirements of transparent, accountable, democratic practices in all areas of governance' and through communities' rights to participation in local development (Williams, 2004, p20).

Stakeholders within the Northern Cape created a multi-stakeholder Asbestos Forum, comprising various government departments[8] to ensure joint consultative decision-making, rather than a fragmented approach to rehabilitation (Minutes of the Asbestos Forum Meeting, 6 November 1997). Community representatives and the CPAA were critically involved in this forum, participating in decision-making and supporting the political process. The Asbestos Forum meetings thus created an 'invited space' characterized by regularity, deliberation and participation (Cornwall, 2004). In so doing, community representatives 'exercised voice' and were expected to 'become empowered' through the process of participation (compare with Gaventa, 2002). However, as Leach et al point out, community participation is particularly challenging when dealing with scientific and technological issues because, on the one hand, 'highly specialised professionalised knowledge and expertise' may restrict participation, while recent scientific controversies have, on the other hand, 'created new demands and opportunities for concerted citizen engagement in decision-making' (2002, p40).

In conjunction with concerted efforts to improve health facilities, the multi-stakeholder forum focused on three main areas of intervention: medical screening, rehabilitation of mining areas and community development (discussed in more detail below). In contrast to the western public understanding of science literature, which sees lay people's scientific knowledge as inadequate and thus limiting their citizen involvement (Michael, 2002, p359)[9] and which assumes a disconnected public, local communities in the Northern Cape were immersed in the issue of asbestos. Community residents accepted the scientific approach as valid and participated in various ways, supporting scientists' attempts to measure systematically the distribution of ARDs or to clean up waste asbestos products in the areas of health and mining rehabilitation. Their participation through collective community action led to broadened understandings of risk which incorporated both scientific and community perspectives.

Medical screening: Measuring the extent of ARDs

Numerous scientific studies, undertaken since the 1920s, have documented the links between asbestos and the diseases referred to above. During the 1950s, pioneering South African medical assessments linked asbestos to mesothelioma. As, however, the asbestos industry, operating in conjunction with the apartheid government, increasingly funded and controlled South African medical research into ARDs, there was a dearth of research publications in the 1960s and 1970s (McCulloch, 2002, 2005). A resurgence of research occurred in the 1980s when anti-apartheid trade unions sought to address occupational health issues (Braun et al, 2003) and in the 1990s when they sought alliances with the new democratic government. At the same time, civil society organizations, such as the CPAA, emerged and challenged the compensation system for ARDs.

Although the classification of ARDs, the degree of illness and the appropriateness of compensation has seldom been value-free and 'objective' (see Waldman, 2007), the underlying political and economic issues have generally been obscured. Furthermore, because the asbestos debate emphasizes complex notions such as minute fibre size, threshold levels, diagnosis of various forms of ARDs and so forth, it has marginalized lay people and ARD sufferers – many of whom in the Northern Cape are illiterate – and discouraged them from engaging in these aspects of the debate (compare with Leach et al, 2002). Community residents did, however, support measures to evaluate risk scientifically. In the 1980s, the CPAA arranged for visiting scientists to collect data and measure asbestos pollution in Prieska.[10] In 1996, the Industrial Health Research Group examined the occupational histories and medical records of asbestos mine workers from Kuruman (IHRC, 1996) and evidence of the long-term risks to miners' health, as well as inadequate health and safety and a lack of information about ARDs and the compensation system.

In May 1997, Dr Ahmon Randeree, a Canadian doctor aware of the Canadian litigations and working in the main Northern Cape hospital in Kimberley, distributed a questionnaire to more than 1000 Prieska residents in order to identify people suffering from respiratory problems. The CPAA supported this research, helping residents to complete questionnaires, informing people when X-rays would be taken and when visiting doctors were doing assessments. CPAA organizers also compiled a database of all residents who were experiencing respiratory problems. Preliminary results suggested that 30 per cent of Prieska's residents were suffering from ARDs, but that only 5 per cent had experienced occupational exposure to asbestos (Asbestos Project Prieska, n.d.; see also Hopley and Richards, 1999). The Northern Cape Department of Health then appealed for help with the medical examinations. Professor Richards from the Gauteng Johannesburg Hospital responded and brought with him a team of specialists to examine patients. This study confirmed the preliminary results suggested by the CPAA and revealed 'a significant number of previously undiagnosed uncompensated cases' of ARDs among Prieska's residents (Hopley and Richards, 1999, p1).

At a community level, there has been widespread support for 'scientific assessments' of the health risks and for medical research. There has also been some attempt, on the part of scientific institutions and personnel, to explain complicated scientific discourse and procedures to the lay public (compare with Michael, 2002). Concerns with health, with the availability of health services and the polluted quality of Prieska's air dominated the scientific/medical surveys (Minutes of the Asbestos Forum Meeting, 6 November 1997). The assessments regarding the distribution of ARDs were both initiated and accepted by the CPAA, by provincial government officials and by locally involved community members. The multi-stakeholder forum thus provided a space in which community residents supported scientific processes and definitions, although I shall later show that the recognition of scientific/medical literature was more ambivalent. Within this multi-stakeholder forum, extensive debate also explored the dangers created by asbestos mining and how environmental rehabilitation should proceed.

Environmental rehabilitation of mines, dumps and mills

The South African government has assumed responsibility for the rehabilitation of the abandoned asbestos mines and dumps and has initiated a reclamation process. Nonetheless, the costs associated with this are enormous and, in 1998, it was estimated that a further R50 million would still be needed (Kisting, 2000). As this estimation exceeds the funding available for rehabilitation, it poses serious problems for local government departments seeking to assist their rural constituencies.

Between 2002 and 2003, the Department of Minerals and Energy (DME) spent R17 million safeguarding the worst Northern Cape mines.[11] In April 2004, the DME reported that it had rehabilitated about 60 per cent – or '111 mines and 578 dumps' – of South Africa's asbestos mines (Venter, 2004).[12] Yet only 45 per cent of the Northern Cape asbestos mines had been treated. In Prieska, the DME reported that rehabilitation, which involved reducing the dump size to an acceptable gradient, covering the polluted area with clean soil, establishing anti-erosion measures and planting vegetation, had decreased the presence of airborne asbestos fibre from 1 fibre per millimetre to 0.01 fibre per millimetre. This has generally been considered an expensive, but ultimately the most practical, solution (Shabangu, 2001).

The CPAA's campaigns to address asbestos pollution emphasized community participation and local development throughout. In most instances this entailed collaboration with scientists, rather than the development of local experts. Nonetheless, Eco-Rehabilitation, a local company, won the tender to rehabilitate the Koegas mine dump, located 40km from Prieska. The owner, a Griquatown resident, employed workers from Prieska, Marydale and Griquatown in order to fulfil government requirements that this be a 'community initiative'. The work involved covering the dump with soil and constructing barriers to ensure that asbestos fibres would no longer contaminate the river. Several of the workers had parents who had been employed on the Koegas mine and all of them had relatives who had contracted ARDs. Eco-Rehabilitation's owner thus commented that the experience was one that 'touched' many of the workers as they realized that ARDs would no longer affect future generations. He himself felt the work to be 'close to my heart, with the knowledge that my ancestors lived at Koegas, and to now be physically involved at the place where they got sick and to be able to fully restore it with state support'. This confidence that ARDs would no longer affect their children was contrasted with concern about asbestos exposure while working on site, although some safety measures were implemented.[13] Although this provides a positive example of a community project with local beneficiaries, the manner in which workers were protected was influenced by socio-cultural interpretations of risk. As discussed below, the workers' focus on visible fibres and the use of face masks may have offered partial protection against microscopic asbestos fibres, but is inadequate from a scientific/medical point of view.

The rehabilitation of mines and mine dumps is, however, only part of the problem. Because mining companies encouraged extensive local use of its by-products, asbestos waste has been found in many everyday contexts in Prieska. Not only is it cemented into the walls and roads, it is in the air – sometimes visible and sometimes invisible – in immeasurable quantities. It is this pervasive and general problem, discussed in the next section, which has provoked the most debate among affected communities.

Compensation, community development and perceived risk

It was common for asbestos waste products to be used in the construction of Northern Cape roads, buildings, temporary school classrooms, and roofs and ceilings (Venter, 2004). The CPAA initially approached companies responsible for supplying asbestos-impregnated products, but without any success. Everite, for example, had supplied Prieska's asbestos cement roofs. When challenged about the inherent danger, Everite claimed that the roofs posed 'little or no additional health risk' (Letter to CPAA, 21 June 1999). Everite's conclusion was based on the scientific evidence of an 'exceptionally low' count of between one and three fibres per eight-hour sample and on 'international literature' which indicated that asbestos cement roofs did not enhance background levels of asbestos fibre. While this assessment follows conventional wisdom that asbestos in situ is best painted to minimize the release of fibres and left undisturbed, daily life does not allow for such clear-cut scenarios. The fact that roofs get broken in violent hailstorms or that people might drill holes to install new gutters and so forth were dismissed as insignificant facts in Everite's scientific calculation of risk. As discussed in Chapter 6, however, Everite's position on the dangers of asbestos has altered radically since 2000, not least because of the actions of the CPAA who, alongside other activists, were able to put asbestos issues on the national political agenda.

In March 2000, the CPAA redirected their efforts to government and wrote to the Northern Cape Premier pointing out that there had been no risk assessments in towns where residents suffered from ARDs, such as Prieska, Maryvale, Griquatown, Boegoeberg, Niekerkshoop and Danielskuil (Letter to the Premier, 20 March 2000). This resulted in only partial success. Although the government envisaged replacing asbestos roofs and toilets 'where practically possible', to date little has been done. In Prieska, it was 2003 before asbestos sheets used in roofing and toilet construction were finally replaced. Although performed as a safety measure against future contraction of ARDs, some people remained in their houses while the roofs were being changed. As one asbestos sufferer commented, 'what difference, [asbestos] is already in your lungs?' In other towns, people have little option but to continue to live in houses built with asbestos products and to take the chance that they and their families might, if they are not already sufferers, contract ARDs. In Griquatown, the CPAA representative campaigned to have the roofs changed, but found it difficult to pinpoint which government department should take responsibility. The Department of Health referred him to the Department of Housing, which in turn referred him back to Health. His experiences are indicative of the high costs involved in environmental rehabilitation, the widespread nature of the problem and the difficulty of persuading government officials, who recognize that asbestos pollution is a problem, that the roofs in a particular town are a priority for redress. In the case of Prieska, the overwhelming scientific evidence of the distribution of ARDs and in particular of mesothelioma, coupled with strong active community mobilization, led to the replacement of asbestos roofs. Other smaller towns do not have the same leverage with government departments, are not seen as being as polluted and have less concerted community action.

The Prieska debates on health and the distribution of ARDs also highlighted the significance of evaluating airborne asbestos pollution. As was the case with the medical screening, there was widespread community support for assessing the presence of airborne asbestos fibres. This, however, was a discussion that took place on two distinct levels – at the level of community concern regarding the dangers of constant exposure and at the level of scientific expertise, disease modelling and equations. At community level, notions of harm and risk concerned visible asbestos fibre and the use of asbestos in construction and infrastructure. At the scientific level, risk was evaluated in terms of wind direction and predicted deposition, microscopic fibres, statistical probabilities and so forth. As pointed out by a health professional based in Prieska, the quality of sampling varied between studies as the seasonal Northern Cape winds affected results (Minutes of the Asbestos Forum Meeting, 6 November 1997). Also at issue has been the analysis of fibre characteristics and correlating these with corresponding degrees of danger, dispersal and risk. Scientific experts disagreed on whether field sites were necessary, on the value of anecdotal evidence and on the accuracy of modelling dispersion patterns (letters from the Institute for Ecological Rehabilitation, 18 June 1999).

In contrast, community residents focused on visible fibres. In Prieska, concerns were frequently voiced: after a cloudburst, several streets 'were exposed to asbestos' (through erosion caused by strong rain); during the seasonal heavy winds, people were able to pick up fibres (notes from the Meeting of the Asbestos Working Group, 24 February 1998); construction work often disturbed asbestos deposits and workers refused to continue until the asbestos had been dealt with (letter from Town Clerk, 9 January 1986). Even sites that had ostensibly been treated were not secure. For example, visits to the rehabilitated Prieska mill revealed asbestos fibre lying exposed on the topsoil. Although resonating with the 'scientific establishment's' debates on fibre deposition and risk, these community concerns about visible fibres were not seriously addressed, despite community involvement in the Asbestos Working Group. Instead they were seen as background to the 'more serious' questions of medical screening and environmental rehabilitation and, although minuted in the Asbestos Forum meetings, did not result in direct action. As suggested by the Prieska mill example, just enough was done to allay people's fears and this primarily involved covering visible asbestos fibres. Where covering the fibres was not possible, such as the case of construction workers exposing asbestos, only material in the immediate vicinity of proposed work has been removed. In other Northern Cape towns, in which the CPAA had a weaker presence and where mobilization around asbestos issues was not as strong, people did not respond to heavy rains, violent wind storms or construction work with concerns about visible asbestos fibres. This reflects how, in Prieska, people's understanding of risk and their scientific knowledge has been enhanced through action and collective participation.

The background of community concerns with visible fibres and with their living conditions has thus occurred in tandem with the emphasizing of medical screening and environmental rehabilitation, and illustrates the widespread and continued faith in science as a means to deal bureaucratically with disease and problems. Ultimately, however, the rehabilitation of mines, the medical assessments of ARDs and of rates of infection, and the limited attempts to address community concerns have not improved the daily experiences of most Northern Cape residents. In the following section, residents' informal assessments of risk are evaluated, with particular attention paid to their understandings of the harm posed by asbestos and to their corresponding ways of coping.

Death in the context of asbestos

Stephens (2002) and Fischer (2005) point out that, in contrast to expert assessments of risk, lay people approach scientific predictions of risk from a socio-cultural perspective. Public concern with risk is thus neither irrational nor based on scientific misunderstanding; but is a vital component of human subjectivity and identity (Wynne, 1995; Lupton, 1999). Michael's concept of 'apprehension' – a process by which lay people 'assess the status of sources of knowledge' and tie these assessments to local identity and culture – shows how people recast and broaden scientific/medical risk in terms of moral judgements and emotional responses (2002, p367). In a process of 'apprehension', and in contrast to what was discussed in the official Asbestos Working Group meetings, informally the people of Prieska and Griquatown emphasized their experiences of damaged bodies, their family relationships and their dependency on ARD compensatory payments for survival. They thus interpreted harm in terms of emotions, bodily integrity and financial responsibility or, to paraphrase Michael, they encountered knowledge through their bodies (2002, p373). Such a perspective offsets the medical and technical ways of describing 'risk' by relocating danger in terms of personal experiences and social relations. For the residents of these towns, what is of concern is not the risks that isolated individuals face, but the wider networks of social relationships and how these are simultaneously threatened by ARDs and sustained through government disability grants and compensation payments for these diseases.

While scientific discussion regarding ARDs has focused on individuals, the Northern Cape is an area where extended families experience widespread and pervasive death from ARDs. It is common for families to watch, and try to assist, many members suffering and dying from ARDs. Emily Julies, for example, witnessed her husband, son, father-in-law and both brothers-in-law dying from ARDs. More than half her husband's extended family (21 out of 34 people) had contracted ARDs, with men and women being equally affected. The widespread presence of asbestos is thus a collective disaster. A Department of Health, Welfare and Environmental Affairs survey revealed that 43 per cent of respondents had ARD sufferers in their family, while 60 per cent of respondents had watched a family member die from an ARD (n.d., p2). Ideas about death thus pervade people's experience of ARDs. The following selection of quotes comes from Northern Cape ARD patients whose experiences have been poignantly portrayed in Hein du Plessis' photographs and accompanying text:

> *I've worked at this hospital for 21 years and I've seen many, many people die from asbestos dust. When you see them gasping for breath you think: 'Lord, what will my hour be like?'* (Anne van Staden, 51, asbestosis patient and hospital employee)

> *I was 12 when I started helping my parents work asbestos at Koegas. Now asbestos will be my end. For the past 23 years I've worked at the hospital and I've seen many people end up in the mortuary. Little did I know that I would also end up here because of asbestos. My father died in 1993 because of the dust and my mother is also suffering because of it. We are five children and three of us also have it.* (Audrey van Schalkwyk, 54, asbestosis patient and mortuary employee)

> *The asbestos is eating me up and it is taking me to my grave. We are dying here.* (Willem Olyn, 63, asbestosis patient)

> *I'm ready to die. But there are 15 people depending on my pension and when I die they will have no income.* (Jan van Staden, 70, asbestosis patient)

The pervasive association of asbestos with ideas about death suggests that an exclusive focus on scientific and technological innovation may not lead to radical improvements in people's lives. Although the actions called for by the CPAA – the rehabilitation of houses, streets, mills and the removal of other sources of asbestos pollution – should create healthier environments and should reduce people's exposure to risk as defined in the scientific sense, community understandings of risk are not necessarily predicated on probability or future anxiety. Instead they are often informed by past experiences, combined with social relations and personal emotions. Indeed, as Michael has argued, people's understanding of science and medicine is intricately interwoven with local cultural dynamics and identities (2002, p362). Wynne's interpretationist approach similarly focuses on understanding scientific knowledge in terms of faith, trustworthiness, credibility and social relationships (Wynne, 1995). It is about understanding how the public experience, interpret and relate to scientific knowledge, about how they assess different claims to knowledge and link these claims to other aspects of their social identity. In the following section, I explore how people cope – and live with – ARDs and asbestos pollution. In so doing, this chapter argues for an understanding of risk that is socially and culturally contextualized. It suggests alternative ways in which people conceptualize the dangers and risks; ways that are not revealed in processes of formal participation, but that inform their coping strategies and that may, ironically, increase their exposure to risk.

Living with asbestos pollution and perceived risk

According to the doctor, I've lived in Prieska too long. (Bettie Jacobs, 75, mesothelioma patient)

In Griquatown and Prieska, people's emic interpretations of their exposure to asbestos are sometimes aligned to, but at other times contradictory to medical and scientific interpretations. Nonetheless, emic understandings of the risks of exposure provide ways of situating danger and of developing locally 'appropriate' responses to these hazards. In so doing, they create possibilities for residents to practise 'good health' and to continue to live in these towns. In contrast to medical assessments and scientific discourses of future probability and exposure to harm, Griquatown residents were most concerned with occupational exposure acquired 20 or 30 years previously, during their employment on asbestos mines and through environmental exposure sustained when they were young and their parents worked on the mines. Even so, Prieska residents acknowledged that the possibility of ARD contraction remained high despite rehabilitation. Although suggestions have been bandied around to shut down Prieska and relocate the people, residents did not take these suggestions seriously. They have chosen, somewhat paradoxically, to remain in the town. Many people born here were emphatic that they would rather die of ARDs than leave Prieska. As a local schoolteacher commented: 'It's their place, this. People are buried here, they grew up here.' Dolf Beukes, an ARD sufferer, explained his attachment to the town as follows: 'I have grown roots here, I was brought up here, born here. It is where my parents were. When I opened my eyes, I was here. I can travel anywhere, but I will always return here. I love this place.'

Similarly, Griquatown residents have returned home after diagnosis of ARDs. As explained by one patient:

> *When I was completely sick, then I came home because I was born here. And all of us … my father is buried here, my mother is buried here, my*

one brother is [buried] here as well, also from the dust because he also
worked on Blouboskuil.

This meant that, on the one hand, ideas about harm and risk were firmly couched in the historical context of mining irresponsibility and callousness towards town residents. As one person commented, 'it is a shame that Cape plc treats our people like this'. On the other hand, these conditions were mitigated by families' historical continuity and ongoing networks of relationships through which ARD sufferers find solace and respite. As suggested above, people affected by ARDs used their pensions to provide for relatives beyond their nuclear families and people turned to their extended kin for support. Families in the Northern Cape towns were connected through their long-term residence in the town, through complicated relationships of consanguinity and affinity that were hard to untangle and trace, through the practice of having *voorkinders*,[14] through extending kinship to people who participated in intimate activities such as childbirth and ritual,[15] and through common experiences of poverty and oppression. It was family members who helped out in times of need, who shared food and clothes with other family members, who drew people into networks of responsibility and obligation and, in so doing, emphasized individuals' value while providing support and solidarity. Prieska residents were characterized as 'caring for each other', as 'helping each other' and as 'being patient with one another'. It was these sentiments that made people feel that they were 'happy', despite knowing that they had been exposed and were likely to contract ARDs. Indeed as Cobb (1976), Dressler (1980) and Kleinman (1995) have argued, social support provides a valuable aid in terms of coping with disease and stress.

Although, as suggested above, most people accepted the scientific interpretations of ARDs, there was considerable disagreement regarding how serious different forms of ARD were (see Waldman, 2007). ARDs are difficult to pin down, both medically and socially. As shown in the previous chapter, one of the primary areas of disagreement concerns diagnosis and degree of severity of the disease. Mossman and Gee (1989), in their review of ARDs, categorize four types of benign disorders stemming from asbestos exposure, namely pleural effusion or fluid on the lungs, pleural plaques, pleural fibrosis and rounded atelectasis. These disorders do not generally produce pain but sufferers do experience shortness of breath and some discomfort. Three serious and debilitating forms of disease have also been linked to asbestos, namely lung cancer, asbestosis and mesothelioma. Asbestosis and lung cancer are primarily occupational hazards, as contracting these diseases is linked to exposure rates. Mesothelioma (a malignant cancer) is unrelated to dosage and trivial exposure can lead to cancer of the abdominal cavity or lung lining. The fact that exposure is followed by an extended latency period (of up to 40 years) before diagnosis, and that the condition is untreatable, heightens the dangers.

In South Africa, people who suffer from ARDs are eligible to receive a state pension if they have more than 40 per cent damage to their lungs (or what is known as second-degree illness in South Africa). This provides a small but steady monthly income. In addition, former employees of the mines are able to receive compensation from the government-run MBOD. The MBOD pays compensation for asbestosis, mesothelioma and lung cancer (Myers, 1981, p241) and distinguishes between 'first-degree' asbestosis or pleural plaques (it used these terms somewhat interchangeably) and 'second-degree' infection (which could comprise asbestos infection plus additional damage to the lungs, either from tuberculosis or from smoking). The apparent similarity between the definitions of asbestosis and pleural plaques raised, for many sufferers, questions regarding medical definitions and legal compensation. People understood that 'second-degree' referred to the increased severity of *mynstof* (literally, mine dust, but also often understood to be asbestosis or any form of ARD) although medically the various forms of ARDs are constructed as different kinds of diseases, rather than as variations in degrees of severity. Pleural plaques are billed as benign and inert and are therefore not seen as being related to other, more severe forms of ARDs or to lead to further damage of the lungs. In Griquatown and Prieska, asbestos sufferers do not, however, draw a distinction between pleural plaques and other forms of ARDs. They do not accept that other asbestos-related complications are not related to the presence of plaques. Their experience of ARDs stresses the interrelated nature of these diseases.

People's emic understandings of ARDs emphasized the destructiveness of all forms of the diseases and they scorned the medical classification of some ARDs as benign or debilitating. From the residents' perspectives, all ARD patients suffered from the physical deterioration of their lungs, and as the following quotes indicate, degrees of seriousness could be conceptualized in terms of 'holes' or absence:

> *I did not even [consider removing my husband's lungs for autopsy and subsequent compensation] because there were no lungs. You see, my husband's lungs looked like that cloth you see there ... you cannot see anything there ... now it's just tendons and [remnants] of the lung that are there, because the asbestos has worked its way through. He looked like that crocheted cloth, there ... he looked like a sieve.* (Emily Julies, Griquatown)

> *I feel like an empty shell that does not have lungs or a heart inside it. Just the other day I commented that it feels as though one of my lungs is already 'light', that's how it feels inside.* (Lena Lucas, Prieska)

The lungs dry out. One day I assisted when they did an autopsy. His lungs were so finished that only the oesophagus remained, there were no lungs, they were disintegrated. (Audrey van Schalkwyk, Prieska)

Thus, in contrast with the medical, scientific view put forward in Asbestos Working Group meetings, which argued that certain forms of ARD were more severe and debilitating than others, emic definitions focused instead on ill-health and on people's ability, or inability, to 'feel' complete. Asbestos was seen to affect not only people's lungs, but to pervade their bodies and their well-being. As one CPAA campaigner described it:

Asbestos settles in people's stomachs. And then those people swell up. You think they have TB [tuberculosis] or some type of cancer and they become thin. Because their stomachs are full of water … It's asbestos lodged in the stomachs.

Medical practitioners working in the Northern Cape saw this as part of a 'gross elaboration' of the dangers of asbestos. They argued that the Northern Cape residents blamed all ill-health on asbestos, but actually tuberculosis and smoking had far greater impacts on their well-being.[16] Nonetheless, in doctors' experiences, so pervasive was the tendency to see asbestos as the cause of ill-health that 'people get very upset when they're told they don't have asbestosis'. In providing their own interpretations of what happened to their lungs and how asbestos fibres affected their bodies, the Northern Cape residents were able to redefine risk and to shape their daily experiences in ways that were understood to provide a degree of safety.

The dangers of inhaling asbestos dust and fibre were readily accepted, with most emphasis placed on, visible fibres, which provoked community action and outrage as they were seen to be fatal. This contradicts scientific evidence which posits that even one-off exposure to microscopic fibres can lead to mesothelioma. CPAA organizers commented that, 'asbestos is not dangerous if it's not visible'. Because of this emphasis on visible fibres, people were willing to work on the rehabilitation of mine dumps wearing only face masks. Similarly, Prieska residents watched their asbestos roofs being removed and cleaned up the mess without requesting protection. In addition, women who had given birth to children were particularly careful about their exposure to asbestos. They believed that cancer of the womb was caused by the necessity of using 'bush toilets' and argued that the wind blew the fibre into their vaginas and their underwear then kept this fibre in place.[17]

These emic beliefs were overly reliant on the presence of visible asbestos fibres and did not stress the dangers of invisible, microscopic fibres. Nonetheless, this position enabled people to continue to live in these rural towns and to cope with their experiences. As Dressler (1980) has argued, cultural factors can serve a 'beneficial or protective function', in that they 'aid individuals in their personal adaptations'. Even though these adaptations may not interpret risks from a scientific/medical perspective, they reduced stress and provided ways to continue living in polluted environments.

Conclusion

The interpretation of risk thus occurs at many levels that are, at times, disparate and contradictory – such as disagreement between medical experts and community residents regarding the severity of different ARDs or the condition of people's lungs – but that can also be complementary – such as the importance of documenting the extent of ARDs in Prieska and the desire to deal with environmental pollution. Scientific interpretations of risk and public understandings of risk are not mutually exclusive. In arguing that ARDs are caused only by visible fibres, residents create a 'safe zone' in which to live. They do not oppose scientific evidence – indeed much of this is widely accepted and supported – yet the information is absorbed and adapted in ways that make everyday life possible, although ironically does not provide protection from invisible and microscopic asbestos fibres. These emic understandings – and residents' attempts to live in asbestos-polluted environments – are susceptible to scientists' and medical experts' claims that most people do not understand the complexity of scientific debates (Flynn, 1999) and that they will 'live with' rather than 'die from' benign ARDs. However, ultimately neither this, nor medical/ scientific understandings of risk – and the actions taken to minimize this – can prevent some people from contracting ARDs. Both the scientific and the emic models have limitations and the 'safe zones' created in both these models do not actually constitute safety for the residents of Griquatown and Prieska. As illustrated in this study of asbestos pollution, local understandings of risk and harm reflect a messy process of convergence and divergence, in which risk and harm are interpreted through the lenses of personal bodily experience and social relationships of dependence. Nonetheless, it is through opportunities for participation and collective action that these perspectives become more closely connected and undergo a process of adaptation, of broadening that brings together the two seemingly disparate conceptualizations of risk.

Yet, as Michael has recently argued, the relationship between science and socio-cultural values is 'discontinuous, fractured and non-linear'. Moreover, the intersection between these two domains is 'uncertain and contingent' (2002, pp370–373). This allows for an accommodation of medical and scientific discourse in the official arena of local governance processes which encourage community participation and for the cultural interpretation of scientific knowledge. However, there are also implications beyond how people participate in policy and in debates that are centred on scientific issues. Although scientific and local knowledge may coexist and intertwine, as is evident in this chapter, the formal processes of participation were framed around medical expertise and knowledge. Despite official representation on the Asbestos Working Group, at no stage did any community residents or CPAA organizers point to possible disjunctures between scientific and informal beliefs regarding ARDs and risk. Was this because local understandings were not perceived as knowledge? Was it because official procedures – through the very process of being formal – define emic understandings and local contextual knowledge as irrelevant? Was it because highly specialized, professional knowledge limited local people's participation possibilities to the realm of supporting scientists? Certainly all of these factors influenced community involvement in Prieska. As Leach et al (2002) argue, providing local people with 'voice' can take place within very restricted frameworks. If local people are to participate more fully in policy processes and debates regarding science, risk and technological hazards, then scientific and medical interpretations have to be correlated with local, situated knowledges. Failure to do so ultimately limits the efficacy of both sets of knowledge – of medical science and of emic values – to protect communities from further risk. While this chapter has focused on local participation and grassroots understandings of ARDs and how these have informed medical debates, the following chapter examines the Indian situation where science and medical opinion regarding asbestos are the preserve of the elite and systematically exclude local understandings of disease and risk. Here, in contrast to the South African experience, evidence of ARDs is constrained by medical science and by the expert opinion which informs government policy.

Notes

1   Isolated legal cases regarding ARDs have involved select individuals claiming from asbestos companies such as Everite. Large-scale cases – against the Griqualand Exploration and Finance Company (GEFCO) and against the former Swiss Eternit Group (that mined at the Kuruman Cape Blue and Danielskuil Cape Blue mines) – have been settled out of court and have resulted in the establishment of the Asbestos Relief Trust and the Kgalagadi Relief Trust.
2   Cobbing is a mining term which refers to the process of breaking ore into small pieces and sorting out the better quality fibres by hand.
3   The Northern Cape experiences high levels of poverty with about half a million people, or 61 per cent of the population, living below the poverty line. This is

slightly above the national average in South Africa, which is 57 per cent of the total population living in poverty (HSRC, 2004). Although official statistics cite only 14 per cent of the Northern Cape population is unemployed (based on people being actively engaged in the search for formal employment), almost 45 per cent of the population is recorded as economically inactive. This latter figure presents a more accurate indication of people's unemployment and poverty (Statistics South Africa, 2003, cited in Northern Cape State of the Environment Report, 2005).

4   Demand for asbestos products remained high in the decades following World War II as expansion occurred throughout the UK. Although some products (such as asbestos yarn and fabric) became less desirable after the 1950s, asbestos insulation board was highly sought after and the market expanded considerably. Cape also extended its production of friction materials, particularly brake linings. As a result, the UK Competition Commission noted in 1973 that: 'Cape is the largest producer of amosite fibres and is responsible for over 90 per cent of total world production. About 30 per cent of Cape's output comes to the United Kingdom, mainly for use in fire insulation board' (1973, pp136–138).

5   The International Labour Organization (ILO) ratified the 'Safety in the Use of Asbestos' convention in 1986 which sought to protect workers and others exposed to it. From the mid-1980s, increasing numbers of European countries banned the mining, production, sale, use and import or export of asbestos (London Hazards Centre, 1995). By the end of 2005 all European Union countries should have banned asbestos (Castleman, 2001).

6   Richard Meeran was working for Leigh Day & Co solicitors and Anthony Coombs for John Pickering and Partners, solicitors, both based in the UK. Both were instrumental in assisting with the legal development of the CPAA and orchestrated the Cape plc court case. Both these UK companies specialized in environmental and abuse claims, with Leigh Day showing a strong interest in the rights of people and the environment in a world of multinational organizations and global responsibility, while John Pickering specialized in work for victims of industrial disease, such as mesothelioma and asbestosis.

7   During the apartheid era, the largest Northern Cape asbestos mines were Koegas, Elandsfontein and Blackridge (A. L. Hall, 1930). These mines closed in the late 1970s because they were worked out or in the 1980s when it became impossible to ignore the associated dangers.

8   Including the Departments of Environmental Affairs, Health, Social Security and Welfare, Minerals and Energy, Water Affairs, Housing, Labour and Nature Conservation.

9   The traditional approach to public understandings of science assumes that scientific literacy is a positive attribute which assists people to become 'better citizens' (Michael, 2002, p359).

10   Although the first study of ARDs (since the early 1960s) was conducted in Prieska in the late 1980s (Reid et al, 1990), this focused on elite white residents and excluded the CPAA's constituency. Nonetheless, the research found high mesothelioma and mortality rates for both men and women. As white women were never employed on asbestos mines, this highlighted the high levels of environmental exposure.

11   South African Government, www.info.gov.za/aboutsa/minerals.htm, accessed 16 June 2005. The Asbestos Working Group, an informal national grouping of concerned people, noted in February 1998 that there were no inventories of rehabilitated and unrehabilitated mine dumps (Letter to the Premier, 20 March 2000). This resulted in the production of an 'Asbestos Rehabilitation Priority Index Database'.

12   Ninety per cent of the asbestos mines in the Limpopo area, 60 per cent of the mines in the North West and 21 per cent of mines in Mpumalanga were rehabilitated.

13   Workers were warned of the dangers, screened for diseases before beginning work and after completion, and provided with safety masks.

14   Children born to young women who were neither married nor in serious, long-term relationships.

15   For example, a woman who helped deliver a baby would be seen as the child's 'grandmother' and would accordingly assume social responsibilities throughout the child's life.

16   Community rejection of the link between disease (particularly cancer) and lifestyles is not uncommon. Balshem (1991) documents how communities living in 'cancer hot spots' refute medical notions of causality, arguing instead that cancer is something everyone is born with and that it can be activated by random acts of fate, rather than controlled through healthy lifestyles.

17   UK studies have shown that women exposed to large doses of asbestos (while making gas masks during World War II) have increased risk of ovarian cancer as the asbestos accumulates in their ovaries (Heller et al, 1996).

Chapter 4

'Show Me the Evidence':
Science and Risk in Indian Asbestos Issues

Introduction

In 2004, and again in 2008, India opposed the inclusion of asbestos on the Prior Informed Consent (PIC) list of the Rotterdam Convention on the basis that India had no evidence that white asbestos, or chrysotile, is dangerous. Inclusion in the PIC list would make it obligatory for asbestos exporters to gain consent from importing countries, thus ensuring that developing countries are not taken advantage of. As suggested by India's opposition to this inclusion, it is not persuaded that asbestos is hazardous.[1] This chapter explores the construction of asbestos risk and disease in India, showing how asbestos has been considered as a tool for Indian economic growth and an instrument of modernization and how, in contrast to the previous chapter on South African mobilization, political awareness of asbestos is shaped by economic interests. The chapter argues that the state's narrow definition of asbestos-related diseases (ARDs) enables it to officially document the lack of ARDs experienced by Indian workers. Unlike the Northern Cape, where grassroots initiatives provided the driving force for social mobilization, this process defines sufferers as politically invisible and inconsequential and, accompanied by the 30-year delay between exposure and the onset of disease, means that there is no obvious constituency to be mobilized.

For many Indians, asbestos is a product of modernization and a means of enhancing economic growth. Asbestos roofs are what poor people desire as symbols of their financial success and modernization. However, for a select few who have been personally exposed to medical debates regarding the effects of asbestos in other countries, asbestos is a lethal carcinogen which should be banned. Behind these opposing understandings lie scientific assessments of asbestos type, fibre size, disease and health risks; technological innovations for the safe use of asbestos; activists' mobilization to create national awareness of the hazards of asbestos; and workers' experiences of ill-health. In order to explore how issues of power, governance and globalization, linked to asbestos, are affecting people in India, the chapter examines the activities of anti-asbestos activists and asks what identities and solidarities emerge from these processes.

Globalization has led many people to contemplate the possibilities of living 'beyond' the state, as members of a global society without adherence to national states and political boundaries. Global citizenship and global governance are seen, in the light of eroding state power and economic decentralization, to provide alternative political controls and freedoms (Heater, 2002; De Sousa Santos and Rodriguez-Garavito, 2005). Underlying this is the assumption that globalization has led to an emerging connection between local and global forces which circumvent national forces and identities (Fox, 2005). De Sousa Santos and Rodriguez-Garavito suggest, for example, that global legal processes are able to protect the local from the negative forces of globalization (2005), through the manner in which local identities, legal processes, socio-cultural values and political norms are increasingly tied to international norms and through the emergence of universal values of global citizenship (Heater, 2002).

Global asbestos production and use reveals stark trends that are linked to processes of globalization, but that reflect economic and power inequalities between countries cooperating for economic gain; almost the entire developed world has banned asbestos as alternatives are widely available (Castleman, 2003), yet the developing world has increased its usage. In relocating commercial activities to developing countries, powerful asbestos mining companies have constructed asbestos as a benign product, asserting that certain kinds of asbestos are less harmful than others and that there are safe working techniques (Castleman, 2001). From this perspective, trade in asbestos should be encouraged by 'appropriate' government policy. These mining companies, based primarily in Canada, have invested in many secondary industries, important trade connections with many developing countries and strong political connections in their own and other national governments. Their vested interests thus stretch across national and global 'spaces', but centre around international regulation processes and national policy processes, as this is where decisions about the banning – or mining, manufacture and sale – of asbestos are reviewed. Canada, for example, uses minimal amounts of asbestos, exporting 96 per cent of this mined product to Asia (Toxics Link, 2002). India no longer actively mines asbestos but imports massive amounts from Canada. In India, asbestos is a thriving industry, supported through policies and import duties that advantage asbestos over other alternatives because the Indian government describes asbestos as a 'gift of God' (Castleman, 2001).[2] The implications of this, for Indian people and for anti-asbestos mobilization, are immense.

Occurring in conjunction with the vast literature on globalization and global governance is the debate on global civil society and identity. Defining global citizenship and its relationship to transnational citizenship is fraught with difficulties. Fox (2005) understands transnational citizenship as a process in which rights-based world views are extended through cross-border civic and political communities. Fox thus distinguishes between state-based citizenship defined by legal political rights and society-based citizenship as collective action and identity-building. At transnational level, this society-based notion of citizenship has also been interpreted as a set of 'values, norms and aspirations' (Batliwala and Brown, 2006, p2). Recognizing the complex and fractured nature of citizenship in all its manifestations, Yuval-Davis points to the multilayered nature of collective identity; she argues that local, national, state, cross- or trans-state and supra-state forms of citizenship coexist and are co-constructed in relation to each other (1999, p119). Such an approach, which brings together the state-based and society-based understandings of transnational citizenship is, as I will show in this chapter, of critical importance.

The network of concerned citizens operating globally and mobilizing against asbestos can be seen as a form of transnational citizenship around which a collective social identity with similar values and aspirations has been formed. Around the world, and coordinated by the work of the International Ban Asbestos Secretariat (IBAS), activists, scientists, grassroots organizations and non-governmental organizations (NGOs) are arguing that the use of asbestos leads to workers and communities suffering from occupational and environmental diseases.[3] This network has sought to pressurize governments to ban asbestos. It has also resulted in the creation of country-specific, anti-asbestos movements such as Ban Asbestos Network India (BANI), an Indian organization that aligns itself with IBAS' transnational aims and values. In this chapter, I integrate Yuval-Davis's multilayered and co-constructed approach with the work of Appadurai who argues that transnational civil society should be seen simultaneously as a project, a process and a space 'in search of an unresolved sociological form' (2006, pxi). This chapter demonstrates how activists seek to make global knowledge more accessible to India's citizens as a project, how the movement against asbestos seeks to build networks and alliances as a processual experience and how these experiences are 'interstitial, overlapping and uneven' (Appadurai, 2006, p2). I argue that these processes, projects and spaces work in ways that are both intimately interconnected and simultaneously enormously distant in India. Geographic and political differences between activists are compressed by globalization processes through the development of transnational citizenship, but class, regional and educational differences between Indian workers and Indian activists are enhanced, because of the ways in which the societal, identity-building (and non-building) aspects of global citizenship occur. Thus, activists based in Delhi identify first and foremost with other transnational citizens rather than with the national constituency for whom they mobilize. Simultaneously, state intervention frames this same constituency as 'inconsequential' and 'invisible'. Ultimately, the invisibility of the workers and villagers operates to reinforce the elusivity of the activists' 'sociological form'. In failing to identify victims, the anti-asbestos movement is unable to translate its vision and global processes into meaning and accountability for ordinary people. As demonstrated in the following sections, this simultaneous compression and expansion of identities is caused by the interrelationships and contradictions between governance and regulation of asbestos, the framing of asbestos knowledge, and the processes of identity-making and citizenship.

Governance and regulation

Globalization, accompanied by mainstream ideas of development which advocate economic growth at all costs, facilitates a cooperative relationship between the capital and the state. In the case of asbestos, the alliance between the Indian government, regulation and global capital seeks to protect neoliberalism and the market economy through a series of inclusions and exclusions. In particular, and as shown in the following section, it does not construct workers as sufferers of ARDs and, in so doing, denies them a form of state-based identity or citizenship defined through legal rights.

India is a democratic republic[4] which has a parliamentary system governed by the Constitution of India since 1950. India also hosts a quarter of the world's poor and about 260 million of its people live below the poverty line. The vast numbers of people in India, coupled with poverty and unemployment, create a context in which a constant and cheap source of labour is available and jobs are fiercely guarded by workers. Simultaneously, however, the Indian government finds that, despite encouraging economic growth as a panacea to development, it is unable to achieve affordable housing, accessible health services and viable livelihoods for the majority. Working in tandem, these conditions make industry – including asbestos cement companies – highly desirable.

The government of India has recognized that economic liberalization has ramifications on workers' health and safety. It set up a working group in the Ministry of Labour for the tenth five-year plan (2002–2007), acknowledging that a 'certain degree of compromise and laxity' had characterized arrangements to date (Joshi et al, 2006, p293). Prior to this acknowledgement, a Supreme Court ruling in the late 1990s had specified that all hazardous and polluting industries be relocated from Delhi to more remote locations (Dupont, 2005). These large-scale industrial projects, designed to benefit the poor, often further marginalize them from access to land, forestry and water (Mohanty and Tandon, 2005, p5). In addition, and as demonstrated below, the minimal occupational safety arrangements and the lack of industrial hygiene surveillance (Joshi et al, 2006), suggests that Indian authorities frequently overlook workers' health and safety.

This laxity and compromise, or what Allen et al refer to as 'institutional fragmentation' (2006, p38), is evident in the policies relating to asbestos in India. Here, policies for asbestos are specified in the arena of international trade that, accompanied by low import duties, encourage the importation of raw asbestos. The delicensing of the asbestos industry in 2003 meant that asbestos can now be imported without a government-approved licence (ANROAV, 2007). Different departments are in charge of different aspects of safety standards, leading to confusion in regulating standards. For example, the Central Pollution Control Board monitors how factories impact on the environment while the Factories Inspectorate deals with affected workers and occupational safety (Acharya, 1989, p590). No specific health policy addresses ARDs (ANROAV, 2007). The diverse uses for raw asbestos, the widespread occurrence of asbestos in buildings and other structures, and the many asbestos concrete products mean that workers are 'covered' by several acts (Joshi et al, 2006).[5] However, while the Factories Act (1948) has special guidelines for workers' protection while handling asbestos, the Building and Other Construction Workers' Welfare Cess Act of 1996 does not acknowledge that handling, sawing or repairing asbestos concrete are dangerous working activities and only recognizes asbestosis as a notifiable occupational disease, ignoring lung cancer and mesothelioma. As Joshi et al (2006) have argued, these acts are often outdated or do not apply to current conditions and create confusion among workers, rather than helping them. Ultimately institutional fragmentation, poor enforcement and lack of compliance mean that these provisions remain ineffective (ANROAV, 2007).

According to a 1993 Indian Supreme Court ruling, industries have to maintain records of their workers' health for a period of 40 years and all industrial workers have to be insured under the Employees' State Insurance (ESI) Act or, if only temporarily employed, under the Workmen's Compensation Act (Toxics Link, 2002). The ESI Act provides for workers' negative occupational health. Nationally, the ESI has a 'sizeable fund', maintained through contributions from employers, employees and the national government. Services are available to workers earning up to Rs6500 per month, which include medical care, sickness benefit, maternity benefit, disablement benefit, dependent benefit, funeral expenses, rehabilitation allowances and vocational rehabilitation training (PRIA, 2004, p35).[6] In Ahmedabad, the Employees' State Insurance Corporation (ESIC) has 14 dispensaries and 189 panel doctors, but its activities are limited. It does not perform regular check-ups on workers, it does not keep records of workers' health or company performance and it does not attempt to address work environments. It thus 'does not feel responsible to address the health problems' of industrial workers (NCDRLD, 1992, p14). Instead, the ESIC waits for workers to approach it for relief assistance. In addition, it is very difficult for sick workers to claim compensation: local ESIC offices refuse to provide the necessary forms, the process is cumbersome and characterized by long delays, doctors employed by the ESIC are not familiar with industrial diseases and workers complain of disrespectful treatment. Thus the ESIC has been seen as 'deliberately obstructive' (NCDRLD, 1992; Sharma, 1998).

Conventional understandings of occupational health envisage a role for the state in which it mediates between industry and workers (an example of this is provided in the previous chapter by the Medical Bureau for Occupational Diseases (MBOD), which compensated thousands of South Africans for ARDs). In India, however, occupational health institutions have tended to support an alliance between business and government. This shift in the nature and expected role of occupational health institutions, from mediation between workers and industry to complete cooperation with industry, further marginalizes workers (Chenoy, 1985). In India, the cement industry is described as the 'heavyweight' of industry and as the 'darling' of the stock exchange. In addition, many government officials and politicians have vested interests in supporting the asbestos industry (Krishna, 2006b). The state is not, however, unaware of asbestos hazards. Linked to the ESIC is the National Institute of Occupational Health (NIOH), which certifies asbestos-related cases and authorizes ESI payments. In 1982 the National Campaign on Dust-Related Lung Diseases (NCDRLD) highlighted workers' experiences and challenged the NIOH on its 'pro-government stance' (discussed further in Chapter 6). The NIOH thus tried to orientate itself to workers' concerns and problems and, as a result, officially documented 2 cases of mesothelioma from an asbestos cement factory in Hyderabad, 2 cases of asbestosis from an Ahmedabad electricity company and 25 cases of workers working in the asbestos jointing and packing industry in Mumbai in 1995.

The Indian government – under pressure to ban asbestos – has conducted research under the auspices of the Directorate General, Factory Advice, Service and Labour Institute (DGFASLI) and held a national seminar in order to sensitize labour secretaries and enforcement officials to the hazards associated with asbestos and other mining processes (DGFASLI, 2006). This seminar focused on India's controlled use of asbestos and did not refer to the global debate regarding the dangers of asbestos. In 2004, the Indian Ministry of Chemicals and Fertilizers commissioned the NIOH to research the health and environmental hazards of chrysotile asbestos in preparation for the Rotterdam Convention[7] and the inclusion of chrysotile on the international PIC list.[8] Inclusion on this list would indicate to governments that all the major international regulatory organizations[9] consider chrysotile to be hazardous. Exporters would have to use proper labelling, include directions on safe handling and inform purchasers of any known restrictions or bans (Liotard, 2006).

India's submission, in line with that of Russia, China, Zimbabwe and others who opposed the inclusion, hinged on the lack of Indian research on asbestos hazards: 'We have studied this issue during the past twelve months with an "open mind"' and 'we are not convinced that the opinion of putting chrysotile on the PIC list is correct ... More time is needed to dwell on this issue', said the Indian representative (Kazan-Allen, 2006, p30). The 2004 NIOH study, co-funded by government and industry, provides justification for India's position at the Rotterdam Convention. The study is based on current workers at asbestos cement factories that have been operating for 10–15 years, completely disregarding the 20–40-year latency period between asbestos exposure and the onset of ARDs. In addition, where evidence of 'impaired lung function' is discovered, the NIOH has been pressurized to identify alternative explanatory factors. The study has not been peer-reviewed by independent scientists, nor have workers or members of the public been given an opportunity to comment (Dutta, 2008a, p2). The collusion between the national government and industry thus creates a set of framings in which it is extremely hard for workers to be identified as victims of ARDs. This alliance has locked India's government into supporting a global understanding of development as something stimulated through economic growth, while simultaneously shedding its responsibility to care for those adversely affected by economic industrialization. These processes of inclusion and exclusion have massive ramifications for activists who seek to mobilize against the use of asbestos. In responding to the state's framings of asbestos as harmless, this mobilization takes the form of knowledge politics.

Knowledge politics and the global asbestos debate

The question 'how dangerous is asbestos?' has long been addressed in the international context. There are two clear protagonists in this global debate. On one side, there are those seeking a global ban on asbestos. This involves activists ranging from IBAS, based in London and keeping track of asbestos developments throughout the world (Kazan-Allen, 2003a), to small localized NGOs working in very specific geographic areas and dealing with immediate community concerns (such as Concerned People Against Asbestos (CPAA) described in the previous chapter). Academics, medical specialists, journalists and lawyers from all over the world are closely aligned with this movement. On the other side are those supporting a continued market for asbestos. Highly visible here is the Chrysotile Institute, an NGO established with funding from the Canadian asbestos mines. As transnational mobilization against asbestos has persuaded governments to ban this product, so asbestos mining companies and organizations such as the Chrysotile Institute have established new markets and buttressed current markets. This process has required repacking scientists' concerns about risk and appropriate regulation and complementing it with new research. Ultimately, both sets of actors have sought to understand the extent of the danger for different types of asbestos fibres.

Widespread international consensus exists regarding the dangers of crocidolite and amosite, also known as amphibole asbestos (McCulloch, 2002; Braun et al, 2003; Landrigan and Soffritti, 2005; Welch, 2007), which are no longer mined or marketed anywhere in the world. This perception of risk has been extended to all forms of asbestos by those mobilizing for asbestos bans (Leigh and Driscoll, 2003; Normark, 2006; Brophy, 2006). For those organizations and people who seek to continue to work with asbestos, the focus has been on white chrysotile asbestos and its fibre composition. As discussed in Chapter 1, it was because of its unique fibre composition – the ability to be woven, flexible and yet strong, durable and fireproof – that asbestos first became famous (McCulloch, 2002). There are substantial differences in the mineral structures and bio-persistence of white, brown and blue asbestos fibres. The chemical composition of chrysotile produces 'white, soft, curly fibres and its fibre bundles have splayed ends and kinks', whereas amosite and crocidolite asbestos have 'needle-like' fibres (Gravelsons et al, 2004).[10] 'Soft' chrysotile fibres are reported to clear from people's lungs within a few months, whereas amphibole asbestos fibres can remain for a year or more (Bernstein et al, 2003, 2005). The inhalation bio-persistence of chrysotile is thus considered by some to be low as it breaks up and decomposes quicker than other asbestos fibres. Bernstein and Hoskins (2006) conclude therefore that low exposures to chrysotile do not pose a risk to health.[11]

This position – that low levels of exposure to chrysotile is safe – is contrasted with the view that fibre consistency is irrelevant and, even if chrysotile has lower bio-persistence, it can still trigger mesothelioma and other ARDs (Egilman, 2003; Landrigan and Soffritti, 2005).[12] Egilman (2003) argues, for instance, that, even if chrysotile fibres are broken down in the lung, they are not expelled from the body and thus still pose a long-term risk. The suggestion that these fibres are also hazardous is supported by the fact that Canadian chrysotile has been, and continues to be, associated with mesothelioma (Landrigan and Soffritti, 2005). LaDou argues that 'actually, on a per-fibre basis, the highest risks have been shown for chrysotile' (2004, p288).

These competing epistemologies have resulted in scientific expertise being divided regarding the dangers of chrysotile asbestos, and this in turn affects the production process. The Chrysotile Institute argues that careful maintenance of a 'practical' threshold level of exposure ensures that there will be no adverse effects' (2007, p4).[13] In the search for safe ways of processing asbestos, considerable emphasis is put on 'controlled use', which includes regulations, good work practices, dust controls, monitoring the work environment, medical surveillance, and education and training for workers. In theory, this 'allows society to benefit from cost-efficient, needed materials used in a responsible safe manner' (Chrysotile Institute, 2007, p6). There is, in addition, some literature which supports the suggestion that chrysotile asbestos can be safely used in production plants. Numerous studies conducted during the 1980s suggest that the statistical probability of risk is low (Thomas et al, 1982; Gardener and Powell, 1986: Hughes et al, 1987). Critiques of this research point, however, to the political and economic contexts in which this research was conducted: to the relationships between the asbestos mining industry and lucratively paid consultants whose research is constructed to exclude negative data. They point also to the fact that 'controlled use' excludes the use of saws and drills on asbestos cement products and the impossibility of controlling in situ asbestos cement in houses, schools or other buildings (Castleman, 2003; Bohme et al, 2005; Egilman and Bohme, 2005; Egilman and Billings, 2005; Welch, 2007). '"Controlled use" of asbestos,' argues Castleman, 'is the asbestos industry's way of referring to business as usual with a false face' (2003, p298). In opposition to the idea of 'controlled use', anti-asbestos movements assert that 'the only "safe use" of asbestos is no use' (Normark, 2006, p7). This activist understanding of asbestos is also grounded in scientific research, but this is research done by independent researchers who are not on the asbestos industry's payroll, and whose work is supported by prestigious medical associations such as the Collegium Ramazzini[14] and disseminated in peer-reviewed scientific publications (see, for example, Landrigan et al, 1999; Nicholson, 2001; Lemen, 2004; Terracini, 2006).

These viewpoints – constructed differently depending on people's position within the debate – are widely circulated. Published articles reflect positively on how asbestos can be safely used (Dunnigan, 1993; Bernstein et al, 2003; Bernstein and Hoskins, 2006; Yarborough, 2006). Other publications challenge this conclusion, questioning the techniques used and putting forward alternative research (Castleman and Lemen, 1998; Egilman and Billings, 2005; Welch, 2007; McCulloch and Tweedale, 2008). Information is also available on websites managed by people on both sides of the debate as they seek to make their version of 'scientific' data more accessible.[15] Many of the articles posted on these websites assess India's response to asbestos debates and to global processes such as the Rotterdam Convention. In addition, a range of trade unions and other advocacy-related organizations have issued publications aimed at reaching policy-related organizations and influencing decision-making. Access to resources, styles of engagement in politics and ability to network influence the dissemination of these competing claims to knowledge.

The Chrysotile Institute is an especially well-resourced NGO as it is funded directly through Canadian asbestos mining. This has enabled it to develop a high profile, employing media and public relations specialists and scientists. It has also been particularly successful in engaging with global organizations – such as the World Health Organization (WHO), the International Labour Organization (ILO)[16] and the International Programme on Chemical Safety (IPCS).[17] For example, an IPCS report, published in 1998, concluded that 'no threshold of exposure has been identified for the carcinogenic risk of chrysotile' (Castleman, 2001, p196). The WHO released two reports in 1997, namely 'Asbestos and Health' and 'Asbestos in Buildings', which have been seen to 'read more like endorsements for why asbestos should continue to be used' (Castleman, 2001, p197). Both reports avoided discussing the dangers of asbestos cement and disregard developing countries' continued use, implying that high exposures are 'a thing of the past' while encouraging hygienic working conditions and restrained, unemotional responses. In the 1990s, the ILO was closely aligned with the asbestos industry. This resulted in people appearing as ILO representatives and projecting pro-mining viewpoints while the ILO failed to take a position on chrysotile asbestos (despite having published stark warnings about asbestos dangers in 1930 and 1974; Castleman, 2001; LaDou, 2004). Over the past 15 years, these and other international organizations' reports on chrysotile asbestos have been widely criticized by the National Institute for Occupational Safety and Health (NIOSH) in the US and by the Collegium Ramazzini for downplaying the dangers and for a lack of scientific objectivity (Castleman, 2001). In addition, in 2007, 81 respected scientists and occupational health professionals wrote to the IPCS, expressing their unease over industrial influence in IPCS reports (Welch, 2007). Inadequate United Nations (UN) funding of the WHO and ILO has increased industry involvement in these organizations and encouraged WHO officials' tendency to be 'lulled into inaction' by conflicting scientific reports (LaDou, 2004, p286).

Until recently, there has been a great deal of confusion over the WHO's position. In October 2006, the WHO issued a policy statement that argued the best way to eliminate ARDs was to stop using asbestos. This statement, which did not distinguish chrysotile from amosite or crocidolite, could not support a 'differentiated approach'. In May 2007, the decision-making body of the WHO, the World Health Assembly (WHA), stated that its 'activities will include global campaigns for [the] elimination of asbestos-related diseases – bearing in mind a differentiated approach to regulating its various forms – in line with relevant international legal instruments and the latest evidence for effective interventions' (Kazan-Allen, 2007). This turnabout from the 2006 position came after critique from the Chrysotile Institute (in February 2007), which found the 'changes in the traditional position of the WHO on the question of health and safety in the use of asbestos' profoundly worrying. The Chrysotile Institute was concerned about a WHO statement at the 2004 Rotterdam Convention that had stressed: 'Chrysotile asbestos is a human carcinogen. No threshold has been identified for the carcinogenic risk of chrysotile. At least 90,000 people die each year from asbestos-related disease' (cited in Chrysotile Institute, 2007, p1). Despite this turnabout, when challenged by IBAS and others on the 2007 differentiated approach, the WHO health and safety specialist emailed that the 'WHO position remains unchanged', supporting the ban on all forms of asbestos in order to ensure the elimination of ARDs. This demonstrates, as LaDou has argued, the attempts of the WHO (and the ILO) to bridge these opposing positions and reach a compromise, but such an approach severely inhibits their ability to address contentious and important public health issues (2004, p288). In the past couple of years, however, the WHO and other organizations have asserted that all asbestos is potentially dangerous and have moved away from the 'controlled use' position. For instance, at the 2008 Rotterdam Convention, the WHO clearly stated that 'chrysotile asbestos is a human carcinogen' and that 'there is no safe level of exposure'. Arguing that the only way to eradicate ARDs is to cease asbestos use throughout the world; the WHO also exhibited safer substitutes in a specially convened workshop (Ruff, 2008; *The Lancet*, 2008). Other international regulatory organizations have also followed suit. For example, the International Agency for Research on Cancer (IARC) met in Lyons, France, in March 2009, where it reviewed this question of different fibre consistency. It found that, although the different forms of asbestos demonstrate different potencies, 'the fundamental conclusion is that all forms of asbestos are "carcinogenic to humans"' (Straif et al, 2009, p454).

The changing global context in which these debates occur, coupled with the under-resourced nature of international development organizations, provide a fertile environment for industry to exercise control in the defining of asbestos risk and hazards. These global regulatory processes on disease epistemology, knowledge and risk are in turn translated into the Indian context. Here, questions about the dangers of asbestos are closely linked to questions of economic development and modernization. The failure of Indian occupational health organizations to mediate between workers and industry, and the alliance between the government and industry, mean that national debates on knowledge politics resonated with – and yet also recast – the international debate.

Knowledge politics and the Indian asbestos debate

The title of this chapter, 'Show me the evidence', comes from an interview with an Indian occupational health researcher. Working for the Society for Participatory Research in Asia (PRIA), an international centre for learning and promotion of participation and democratic governance, he participated in the 1982 National Campaign on Dust-Related Lung Diseases (NCDRLD), presenting a paper on asbestos hazards at a manufacturers' conference. He was immediately asked to withdraw his paper or face legal charges. A representative from the manufacturers' conference said 'show me the evidence ... show me an Indian study saying that it's harmful'.

Almost 30 years later, the need to show evidence continues to be a significant part of the discourse used by the asbestos industry and the Indian government. As demonstrated above, it was the reason for India's rejection of the Rotterdam Convention in 2004 and again in 2008. Linked to this need to show Indian evidence is a strong sense that the Indian context is different. 'In the Indian context, some things have no relevance,' said the chairman of the (Indian) Asbestos Institute,[18] who added that India had learnt from western countries' mistakes and could therefore handle asbestos differently:

> *Today, each country is responsible for its own citizens. We have a large medical profession and a lot of scientists doing studies. It should be left to the country – to India – to decide what is good for me ... the government is convinced that the use of white asbestos does not pose any health risk. When we find that this is harmful, then we will stop using it ... It is not out of ignorance, it is out of knowledge that we are using it.*

This emphasis on Indian evidence to explain the Indian context is a way of countering growing international consensus and opposing moves towards a global ban. In fact, a double-edged process is happening whereby, on the one hand, the government places the responsibility on anti-asbestos campaigners such as BANI[19] to demonstrate that asbestos is dangerous while exonerating those that import, produce and market it from this responsibility; while on the other hand, the state simultaneously defines ARDs out of the official system, making it impossible to prove the hazards of asbestos.

The definition of ARDs in India is very narrow, focusing primarily on asbestosis, although some official statutes also include lung cancer and mesothelioma. The official government report carried out in preparation for the 2004 Rotterdam Convention argued that it had conducted a 'multi-disciplinary National project' in which 702 workers were examined, but found 'no established case suggestive of asbestosis' (DGFASLI, 2004, p3). Another study released in 2005 argued that 'large and medium scale industries in the organised sector' had taken 'all necessary safety and environmental control measures and are maintaining excellent hygiene, health and safety conditions' (DGFASLI, 2005, p7).[20] It recorded airborne asbestos fibres well below the legal threshold limit of $0.1f/cm^3$ in all these industries and used this to sidestep the question of whether this threshold – that is considerably higher than that permitted in other countries – is safe. Interim reports on a recent study, being carried out by the

NIOH, suggest that it will reach similar conclusions (Dutta, 2008a). This means that very few cases of asbestosis, lung cancer and mesothelioma are diagnosed in India. In 2005, it was reported that less than 30 workers had been compensated by the Workmen's Compensation and ESI Acts; while occupational lung cancer had never been officially diagnosed and compensated (Murlidhar and Kanhere, 2005). Such figures support the government's argument that Indian asbestos is safe and that controlled methods of production protect workers.

Despite these and other studies, there is a general sense that there is very little Indian research demonstrating the dangers of asbestos. This perception, reiterated by actors across the political spectrum, does not, however, stand up to scrutiny. A growing body of Indian scientific research points to the hazards of asbestos and to the presence of asbestosis and other related diseases (Nath, 2000; Ramanathan and Subramanian, 2001; Murlidhar and Kanhere, 2005). As, however, this research does not focus on epidemiological debates regarding fibre size and bio-persistence, and instead examines the political and economic contexts in which workers are positioned, it is easily dismissed. For example, Murlidhar and Kanhere (2005) reported that of an estimated 100,000 workers exposed to asbestos in India, fewer than 30 had been compensated for ARDs. Refuting the suggestion that Indian asbestos is safe, they turned their attention to the political reasons why so few cases have been recognized. These include workers' difficulty in obtaining medical certifications from their employers, deliberate misdiagnosis, management's control over workers, over the factory space and over the results of any medical surveillance and finally, what they term, the 'healthy worker effect'.[21] Similarly, Nath's work demonstrates the peripatetic nature of construction workers, and he argues that their temporary, migratory lifestyle means they often have no written record of employment (2000, p316). As far as exposure to asbestos fibres and official records of ARDs go, these workers are simply invisible. As far as legal state-based notions of citizenship go, these workers are denied the basic right to a healthy working environment.

India's failure to prioritize occupational health is accompanied by a severe shortage of trained medical personnel (Joshi et al, 2006). India's first occupational health course aimed at medics was launched, amid much scepticism from the medical establishment, in 2005. As Professor Joshi was unable to fill the course, he approached the government for help. Dr Deepak Agarwal was, at the time, working as a government doctor and, as the youngest and most subordinate, he was made to attend. Prior to attending the course, he was vaguely aware of diseases such as silicosis ('just as a disease'), but unaware that such diseases were preventable and untreatable. His lack of experience in occupational health issues is common in medical circles. Despite the fact that this course, and others similar to it, are now more widely available, the medical community has very little sensitivity and professional training in occupational health. Doctors are unaware of the need to look for asbestos-related and other occupational diseases and often treat the symptoms as chronic bronchitis or tuberculosis (Murlidhar and Kanhere, 2005). Professor Joshi recounted one extreme case where a patient with mesothelioma was being treated for pneumonia. In exceptional circumstances, when doctors come face-to-face with asbestos-related illnesses, they are reluctant to get involved in the 'medico-legal aspects' (PRIA, 2004, p4).

In the Indian state's view, in exceptional circumstances where workers do contract occupational diseases, they are eligible for ESI. However, because asbestos is often produced in special hazardous zones which allow companies to exempt themselves from ESIC, workers are often not eligible for compensation. This lack of eligibility further reduces the number of recorded instances of ARDs and, in turn, reinforces the view that Indian techniques for working with asbestos are safe. In addition to a legislative environment that defines asbestos diseases in limited terms, doctors' inability to recognize occupational health diseases and bureaucratic procedures making it difficult for workers to claim compensation, workers are themselves ignorant of these diseases and act in ways which further undermine the visibility of asbestos victims. It is in this context that the NIOH is able to record – as a victory for workers – the identification of mesothelioma and asbestosis cases in Hyderabad and Mumbai. These cases were also significant because they showed that 'longterm exposure to any type of asbestos can lead to development of asbestosis, lung cancer and mesothelioma' (Krishna, 2006b, p25) and also because these results were presented in parliament.

Knowledge politics in India draw on – and restructure – global politics about asbestos risks and dangers. India's quest to facilitate economic growth and to develop through economic growth means that particular kinds of knowledges come to be seen as more pertinent: Indian examples and statistics are required rather than drawing on the experience of industrialized nations; epidemiological understandings are elevated while socio-political and economic analyses are dismissed. Ultimately, both the framing of the governance and regulation of asbestos and the competing interpretations of scientific knowledge about asbestos are also influenced by – and influencing – activists' mobilization and identities.

Activists' mobilization and identity

State decisions regarding regulation, threshold levels, identification of disease and so forth are, as Yarborough (2006) suggests, political decisions which reflect public policies rather than scientific application into epidemiological studies. In the case of India, the government has chosen to align itself with the asbestos industry. As demonstrated above, it therefore defines ARDs in very narrow ways that perpetuates the idea that these diseases are not prevalent. These decisions, and associated government policy processes, affect a wide range of people and influence many people's identity and experience of citizenship both by acknowledging some rights and freedoms and by demarcating the limits of participation.

Acting as citizens mobilizing for certain rights requires a sense of agency. People can, through mobilization and participation, shape their citizenship by claiming new rights, demanding accountability from the state – or supra-state organizations – and limiting state dominance (Mohanty and Tandon, 2005). Grassroots struggle also involves building a collective identity, as people connect with each other and realize the widespread nature of the problem. Citizenship is thus about making connections with other activists experiencing similar problems or interested in similar issues; it is about self-identity and social recognition. Thus, how people conceive of themselves, how they are conceived of by others and who they relate to, affect their mobilization struggle and their ability to act (Kabeer, 2002). As Appadurai has argued, the transnational organizing potential of social movements is linked to 'their capacity to recognize and identify each other, across numerous boundaries of language, history, strategy, and location' (2006, pxi). Paradoxically, the nature of people's mobilization strategies will, in turn, also affect their identities.

Mobilization for citizenship rights involves certain contradictions. It stems out of, but in turn shapes, people's identities. In addition, collective struggle is often necessary to achieve individual rights (Fox, 2005). Globalization enhances these contradictions by adding layers of complexity: 'world' citizenship addresses environmental and other socio-political problems by directing struggles towards an 'ultimate' global level of rights. This is possible because, as Held and McGrew argue: 'Political decisions and actions in one part of the world can rapidly generate worldwide ramifications. Sites of political action and/or decision-making can become linked through rapid communication into complex networks of political interaction' (2002, p5). The disconnection between place, social situation and politics is thus another contradiction inherent in citizenship mobilization. In addition, and linked to this disconnection, is the porosity of national boundaries. Globalization has meant that national communities can no longer make exclusive decisions and policies for national citizens. Rather, the 'pursuit of effective government and the accountability of political power is no longer coterminous with a limited national territory' (Held and McGrew, 2002, p7). National government is, in effect, sandwiched between supra-state organizations such as the UN, regional organizations, transnational civil society and corporate business, local or city government, NGOs and communities. Multifaceted and intricate networks of governance are set up, involving state, supra-state, sub-state and private-sector actors, yet extend well beyond the state (Scholte, 2000). These contradictions raise questions about how mobilization happens and towards whom the struggle should be directed. What are the most 'appropriate targets for influence' (Batliwala and Brown, 2006)? Should mobilization focus on claiming rights across borders, generating a transnational or global identity and citizenship, or will grassroots struggle and challenging the national government provide the best results?

Transnational 'space-based' mobilization

Indian activists working on anti-asbestos campaigns take on a daunting challenge. BANI is a very small organization consisting of civil society groups, trade unions and human rights groups. Being primarily urban-based, it is an alliance of scientists, doctors, public health researchers, trade unions, activists and civil society groups which 'condemns the government's continued pro-industry bias and lack of concern for the asbestos-injured' (Krishna, 2006b, p25) and demands an immediate ban on all uses of asbestos in India.[22] BANI has succeeded in drawing society's attention to hazardous and toxic products (Krishna, n.d.), in rallying medical experts to demand the phasing out of asbestos (Misra, 2003), in filing complaints to the National Human Rights Commission, in pressurizing the government to initiate studies and in partially reorientating the NIOH towards workers' concerns and needs.

Scholte (2002) argues that six factors are crucial to the success or failure of civil society organizations that seek to challenge the state. These are the availability of resources, access to networks, the attitudes of officials, the receptivity of the mass media, the political culture of a given country and the accountability of the groups themselves. An examination of BANI's mobilization experience shows that the bulk of these factors work in ways that inhibit their mobilization. However, because of the way globalization has shaped relationships and identities, the accountability of the groups and the access to networks have more ambiguous consequences, working both to support BANI's activism and, simultaneously, to undermine their activities. As a fledging organization founded in 2002, BANI has faced a number of difficulties in its operations. Some of these were, as Scholte (2002) predicted for organizations in general, linked to the negative responses they received from the media and from civil society. At one point, BANI had been able to influence the government through the media, but the media has subsequently aligned itself with government and industry. After the publication of a spate of BANI-influenced articles on the dangers of asbestos, the asbestos cement industry published several advertisements declaring asbestos safe. This resulted in mainstream media refusing to publish BANI articles while public tirades have labelled BANI as anti-development, elitist and not caring about the needs of the poor.

In keeping with Scholte's criteria for success, BANI works in a context where government officials' attitude towards them is very negative. For example, Professor Joshi, a prominent member of BANI, was threatened with legal action by the asbestos companies. In conjunction with this, government officials suspended him from his work at a government national health institution and remained unsupportive when he faced threats to his professional integrity. This lack of government support is linked to the political culture of India which makes it harder for activists to mobilize. Although India is an established democracy, this does not necessarily facilitate anti-asbestos mobilization. Instead, the government establishes spaces for participation, but these are bound by rules, codes for representation and inclusion which ensure that the government remains dominant (Mohanty and Tandon, 2005). In the anti-asbestos movement, there is little room for negotiation. In contrast to many other social movements, BANI is seeking neither to collaborate nor to establish a consensus. It emphatically states that there is no room for negotiation: 'We can't say we can live with asbestos. We say what we feel needs to be said, we put ourselves in difficult positions but what's the point of being here, what's the point of doing things half-way? We have to be reasonable, but we are also completely clear [that asbestos must be banned].'

As observed by Mohanty and Tandon (2005), mobilization and making demands on the state is never easy; not even when the state is committed to progressive, democratic and participatory procedures. This is because society's relationship with the state has constantly to change and develop in order to cope with a state which makes progressive policies, but continues to enable divisive politics; claims to privilege scientific knowledge, but does not necessarily do so; and stage-manages citizen participation. This means that, even under ideal circumstances where participation is encouraged by the state, people experience difficulty determining their role and identity as citizens because the elements of that identity are constantly shifting in response to the machinations of the state. BANI, it seems, has responded to this challenge by not defining itself in assertive, definitive ways. Rather, as BANI members point out, it ebbs and flows, its strengths come and go. By not having a clear presence or, in the words of one activist, by being a 'silent network', BANI is also harder to pinpoint and this makes it harder for the industry to try and persuade members to change allegiances. BANI's sense of purpose is, as Falk would argue, aspirational, based on a conviction that it is necessary to 'force the impossible to happen' (1994, p132; see also Appadurai, 2006). BANI aims to achieve an asbestos ban, but also to make its message heard: 'We are not going away. It doesn't matter if we win or lose, it needs to be said and we must say it.' This means that, at the national level, BANI and the government are locked in direct conflict. The government is determined that asbestos can be monitored and controlled. It has, as one BANI member put it, 'become closed to listening about asbestos'.

India's policy of economic growth, and its accompanying lack of concern for workers, has also provided industry with considerable autonomy which enables factories to protect themselves against outside scrutiny. Access to factories and other industrial sites is strictly controlled, making it very hard for BANI to compare conditions, develop widespread awareness or to share experiences of working conditions (PRIA, 2004, p4). Asbestos companies are not required to disseminate publicly their health and safety records. Structural economic conditions further decrease BANI's mobilizing opportunities. Despite the emphasis on economic growth, India remains a country where millions of people are unemployed, housed in slums, sleeping on the streets and surviving below the poverty line. This places massive incentives for people to want to keep their jobs, regardless of the employment conditions. Workers and trade unions thus tend to prioritize employment over health. Thus, although some trade unions are interested in an asbestos ban, many prefer to emphasize safety devices and diagnostic tools. As Mr Singh said, 'so we do collaborate to create awareness and put pressure on the government, but our primary concern is not the banning of asbestos'. It is thus clear that, despite being a democracy which emphasizes participation and the importance of knowledge, India's political culture and the accompanying socio-economic conditions impedes BANI's attempts at mobilization.

Finally, Scholte (2002) identifies the accountability of the group and the nature of its networks as critical factors influencing mobilization. In the case of BANI, these two factors are interrelated and work in ambivalent ways which both undermine and support BANI's mobilization strategies. In terms of accountability, BANI operates in a context where the political constituency is missing because ARDs are narrowly defined and there is no official asbestos disease registry. This lack of evidence is keenly felt by BANI, which needs to identify victims as a source of mobilization. As one member of BANI explained, it has to be able to produce hard evidence of what is happening in India, it has to be able to show the health effects if this is to be a driver for change (discussed more in Chapter 6). But the current failure to identify and record victims means that there is not much political agency to push this forward. Another BANI member phrased this in terms of mass action and mass awareness. He argued that, as long as no victims were identified and people were not being seen to be affected by asbestos, workers remained 'bogged down by bread and butter issues' and preferred not to 'rage a struggle'.

BANI's strategy is very different to previous attempts to address occupational health in India. The first occupational health struggle run by PRIA in the early 1980s was the National Lung Campaign which prioritized workers and targeted these people in terms of its mobilization strategy. It focused on the 'conscientization and leadership building of worker activists' (PRIA, 2004, p1). Campaigners thus sought to make workers ask questions about accountability and citizenship rights. In response to this, the government was considered to be the duty-bearer and claims were made directly on the state. The lung campaign did have some successes: the NIOH was forced to adopt a more uncomfortable mediating position between the government and workers and to be more accountable to workers; although, as we shall later see, it has not always succeeded in representing workers' interests. It was also able to stimulate the creation of several grassroots organizations addressing occupational health (see Chapter 6). These grassroots organizations, started by workers who had personal experience of the work environment and of their fellow workers' deterioration in health, have demonstrated their political sustainability, successfully challenging the government on a wide range of occupational health issues such as sewerage workers' exposure to hazardous conditions and cotton workers' experience of byssinosis.

BANI's origins stem from the National Lung Campaign as well as from the more recent possibilities created by globalization. During the National Lung Campaign, Barry Castleman, an environmentalist and researcher specializing in asbestos hazards, who has links with the IBAS and the Collegium Ramazzini, visited PRIA in New Delhi. This established the transnational links out of which BANI ultimately emerged. Most BANI representatives interviewed had begun to question the safety of asbestos because of their international connections. For example, Professor Joshi was influenced by medical colleagues at the Mount Sinai Irving J. Selikoff Center for Occupational and Environmental Medicine in New York, whereas journalists followed European and American media debates. BANI's approach, its characterization of itself as a silent network, is thus a reflection of its origins. This approach, while effective in terms of remaining 'out of the reach' of the state, makes it difficult for BANI to maintain an active presence in India. This ephemeral presence leads many people to argue that 'the edge has been lost' and that BANI is not sufficiently proactive. BANI does not have an overt presence, does not appear to be organizing meetings and it is no longer able to make as much use of the media as it would like. In contrast to the CPAA described in the previous chapter, BANI does not have an office or a logo or a public relations person. While this lack of presence is interpreted by many people as evidence of BANI's lack of political muscle, BANI sees these as positive attributes providing a shield for political activity: 'We are able to do things that are not on our website, not in the public zone. There is no name associated with it, but it's very important for the network.' Electronic communication makes this form of coalition and mobilization possible: 'Rather than be represented by a building that people enter, these actors may be located on electronic networks and exist as "virtual communities" that have no precise physical address' (Rosenau, 2002, p58), but provides a critical medium of engagement. For example, in the case of the French ship, the *Clemenceau*, 'BANI was used, rather than there being action within BANI'. It was primarily international actors within IBAS who organized protests in industrialized countries and effectively mobilized in BANI's name.

This is not to say that there is no sense of a BANI presence or that BANI is not proactive in mobilizing. Those few BANI members who take overt, public stands are protected by their international links and networks, which are primarily facilitated through IBAS. Landrigan and Soffritti (2005) point out that this support is critical as Indian scientists and other involved persons have received death threats and attacks on their professional careers. When, as discussed above, Professor Joshi was threatened with legal action and suspended from his government research post, his international links and a 'virtual network of international support' sustained him as he refused to withdraw his anti-asbestos statements or to terminate his activist activities (Castleman, 2002, p3; Landrigan and Soffritti, 2004). These international connections also provide a vital source of inspiration for many Delhi-based activists. These networks are critical avenues through which BANI members forge a shared identity with other transnational actors. Fox (2005) has suggested that the analogous roles played by activists located in different parts of the world create a shared status of 'counterparts'. In addition to the explicit desire to ban asbestos, BANI activists living in New Delhi and other Indian cities share broad ideological and political values with other transnational anti-asbestos activists, including high levels of education as doctors, journalists and lawyers from India connect up with others in the UK, US and elsewhere; the pursuit of middle-class standards and democratic values; the interconnectedness of their lives through computer technology; and so forth. As Fox (2005) has indicated, the sharing of these extra-mural values enhances the shared collective identity.

BANI's identification with international anti-asbestos campaigners, rather than with workers, is evident in some of its recent activities. In campaigning against the dismantling of the *Blue Lady*,[23] an asbestos-laden ship exported to India for decommissioning and dismantling, BANI focused on establishing transnational networks, setting up meetings and bringing together various parties to put pressure on the state. In May 2006, BANI requested that the Supreme Court ensure the *Blue Lady*'s compliance with both international and Indian law. Activists pointed out that the importing of asbestos waste was banned under the Hazardous Waste Rules, 2003, and under the Indian Environment Protection Act; therefore dismantling the *Blue Lady* was a violation of the Basel Convention (Ganguly, 2007).[24] These concerns were, however, overruled by the Supreme Court in June 2006 which, despite activists' protests and media outrage, permitted the ship to anchor offshore at Western Alang Shipyard.[25] Permission to anchor was granted on 'humanitarian' grounds by the Supreme Court while the Pollution Control Board reversed an earlier decision and allowed the importation of the ship.[26] The Supreme Court then constituted a Technical Committee on Shipbreaking to examine the health risks associated with dismantling. During this investigation, BANI representatives focused on a legal battle with the state. They pointed out: 'The court is perverting the meaning of environment and using it to nullify the needs of local people or to justify things that they are doing.' BANI made sure that its lawyers were present at all the court hearings to witness the Supreme Court's actions; it planned ongoing research into the dangers of asbestos and an international campaign with simultaneous demonstrations in Germany, Norway and India to reveal the double standards of countries that sent ships to India without prior decontamination. This 'boomerang approach' (Keck and Sikkink, 1998) or 'stretching' of politics (Held and McGrew, 2002) has the advantage of using transnational advocacy as a means of outflanking the Indian government and putting international pressure on a domestic issue.

Ultimately, this campaign was unsuccessful and, despite unofficial estimates that the *Blue Lady* had as much as 1200 tons of asbestos and other carcinogens, the Technical Committee of Shipbreaking reported that it did not find 'anything harmful for people and environment in the dismantling of the *Blue Lady*' (Gujarat Global News Network, 11 September 2007). The Supreme Court decided, in September 2007, that the ship could be dismantled at Alang, accepting the Technical Committee's conclusions that a dismantling plan would not cause undue harm and exposure and prioritizing development concerns over health and environmental considerations (Venkatesan, 2007). Disappointed BANI members pointed to the various discrepancies between Indian Law, Supreme Court directives and actual practice (Krishna, 2007b).

What BANI did not do in this campaign was to engage directly with the workers in Alang or anyone else experiencing occupational exposure to asbestos despite the fact that villagers in the vicinity of Alang complained and the Panchayat of Talaja District had filed an application on behalf of all 30,000 people living within 25km of Alang (Ganguly, 2007). Instead, BANI's campaigns were aimed at an international audience, orchestrated to take place simultaneously in Norway, Germany and India. This was intended to facilitate 'maximum' impact in a manner made possible through globalizing processes of media reporting, internet, email and other forms of information technology. As this example makes evident, ultimately BANI is fundamentally a cosmopolitan movement which exists in a 'now you see it, now you don't' modus operandi. This style of moblization is intimately connected to international processes and international anti-asbestos movements. It is possible to trace a correlation between the waxing and waning of BANI's activities and international ban-asbestos activism. BANI's primary interaction – and target – is the government of India. Although many of its members are based in cities in other states, it is not a movement which has a strong domestic presence in the rest of India. This is in keeping with much of the literature on social movements, mobilization and globalization (Laclau and Mouffe, 1985; Touraine, 1988; Escobar and Alvarez, 1992; Falk, 1994; Keck and Sikkink, 1998; Price, 1998; Sikkink, 1998; Edelman, 2001), which emphasizes transnational civil society and NGO participation at the expense of local, small-scale public action that supports these global processes. As Escobar has argued, 'the concern with space has led to a marginalization of place that has consequences for how we think about culture, nature, development and the like' (1999, p292). BANI's inspiration came from global networks, processes and influences and it challenged the manner in which national processes systematically defined the use of asbestos as safe, failing to acknowledge the threats to workers' health. In its association with transnational 'spaces', BANI has operated in ways that distance it from grassroots mobilization around asbestos issues in India. This is, as Appadurai has argued, partly a feature of globalization itself in which the 'discourse of globalisation is itself growing dangerously dispersed, with the language of epistemic communities, the discourse of states and inter-state fora, and the everyday undertanding of global forces by the poor growing steadily apart' (2000, p2). He suggests that the poor and their representatives are thus increasingly isolated from national discourses about globalization as well as from debates and policy discourses regarding trade, labour, environment and disease. As the following section demonstrates, grassroots mobilization takes a very different format in Gujarat, despite similar political processes.

Grassroots 'place-based' mobilization

Van Steenbergen (1994) is not unique in arguing that citizenship is a dynamic concept which requires some form of identity-building and participation in order to realize this status (see also Scholte, 2002; Lipschutz, 2004; Batliwala and Brown, 2006). He argues that because the toxic movement is about workers standing up for their rights, grassroots social activism is vital. For van Steenbergen, citizenship involves rights, entitlements, duties, obligations and responsibilities and this is achieved through '"being part of" as well as to being active in and fully responsible for' (1994, p164). How successful mobilization strategies are in claiming these rights and creating active processes of citizenship is, as we have seen in the case of BANI above, linked to the availability of resources, access to networks, mass media receptivity, the accountability of groups, the political culture of the country and officials' attitudes (Scholte, 2002).

In Gujarat, in parallel with national policy processes, occupational health and workers' rights do not have a high profile, as they are undermined by both Gujarat's political culture and officials' attitudes. Here, industrial investment is highly sought after by the Gujarat State. After the government of India introduced the New Industrial Policy in 1991 (aimed at creating economic reform in the industrial sector), the state of Gujarat aggressively promoted and facilitated new industrial development through concessions and subsidies (Hirway and Mahadevia, 2004). In 2007, Gujarat had nine manufacturing units or factories that used asbestos in the production of asbestos cement sheets and pipes. Three of these factories use dry processing, while the remaining six use wet processing. The Director of Industrial Safety and Health and the Chief Inspector of Inspection (construction) for the state of Gujarat, Mr Gupta, claims that 40 years of factory medical records – which include pulmonary function tests, blood and urine tests and chest X-rays – demonstrate the success of wet processing techniques and personal protective clothing: 'Because of all of this, we have had no cases [of ARDs] in the last three years.' Based on these records, the Department of Industrial Health and Safety has identified only two cases of asbestosis in 2002/2003.[27] Despite the fact that industries monitor dust collection systems and workers' health, the state of Gujarat is proud of its safety record. Gujarat is – in the view of the Certifying Surgeon – 'number one in terms of safety'.

In Gujarat, officials' disregard for workers' health is further reinforced by religious viewpoints. Mr Gupta firmly believes that occupational health levels are also influenced by God. Gujarat has had no major industrial accidents, because 'God is here in Gujarat, who takes care of all these things'. To date, he argues, disaster management has not included spirituality and this is a 'missing dimension'. Spirituality makes you, as a worker, 'more aware of your soul who is running your body and taking care of your own health'. If workers and industry 'believe in God, trust in God and work with God, then production, health and safety would be in a good condition'. Such an understanding of spirituality places the onus on the workers to be 'well aware' and to ensure that accidents do not happen. Telepathy should enable workers to sense that something is going to happen and to take preventative action. Workers should thus strive to achieve a mental balance and supreme energy. Termed 'disaster management with a new and unique approach', this approach ensures that workers are held responsible for their own illness.

Overall, there is an official denial of workers' health issues. Bharuch houses one of the biggest industrial units in India and Everest's asbestos factory is based here. With the help of Mr Malik, the Deputy Director for the Department of Industrial Safety and Health in Bharuch, I was able to visit this factory. Everest's factory manager had the same understanding of asbestos as that of government officials described above, saying 'if asbestos is safely produced, then there are no problems'. Over the past 15 years, he reported no sick workers and no complaints. Demonstrating his supreme confidence in asbestos, he took us into the factory – where several young bare-chested men were weaving strings of asbestos yarn into ropes without gloves or masks – and fetched a handful of asbestos fibres for us to see. Throwing this onto the ground, he reached into his pocket with the same (unwashed) hand and passed us all cotton masks for our mouths. Alongside us, the storage drum for the glycerine-based wetting agent was empty and clearly not in use. This contravened the Gujarat Factories Rules, 1963,[28] but concerned neither the factory manager nor Mr Malik, the official state factory inspector.

However, unlike many other state officials interviewed, Mr Malik believes that asbestos is dangerous. He is acutely aware of the politicized nature of asbestos production and that developed countries 'get themselves safe while sending hazardous industries to developing countries'. As a government official, he thus has an unusually sophisticated understanding of the health and politics surrounding asbestos use in India and internationally. Officially, he is responsible for supervising its use in Bharuch, but he does not have the scientific equipment to monitor airborne fibres. At Everest, however, he condoned the obvious flouting of safety regulations. This lack of concern is related to India's 'unofficial' policies. The flouting of regulation is a widespread and well-recognized feature of India's industrial development (Dupont, 2005; Joshi et al, 2006). 'In the Indian context, some things have no relevance,' said the Chairman of the Asbestos Institute, during an interview in Delhi. He continued:

> *The mistakes [made by European countries] stem from the use of blue and brown asbestos which were used during the period of ignorance with high concentrations. But now levels of workplace exposure are controlled. There were no precautions and people used the material very freely. Now people understand and precautions have been taken. India's environmental pollution [control] is very advanced and based on international levels. No asbestos is seen in the entire factory, no one touches it.*

These comments are, however, far removed from reality. As this example shows, in practice, workers are exposed to asbestos and management shows a casual disregard for appropriate safety measures. Indeed, as Mahesh Banerjee, an NGO worker, pointed out, monitoring is, in effect, a means of protecting the industry. For example, workers' complaints to the Pollution Control Board are seldom investigated as officials accept bribes from factory managers. Even if a factory is 'officially closed', a promissory note explaining how the problem will be addressed is sufficient to revoke the closing order. In practice, work continues as normal.

Companies which are considered to be polluting have to submit Environmental Audits every six years. But it is the act of monitoring – rather than the content of the reports – that has significance. Once submitted, these Environmental Audits are not scrutinized or analysed. This enables the Gujarat Pollution Control Board to comment that, despite receiving more than 700 Environmental Audits, there have been no irregularities and no need for follow-up action. The chairman of the Asbestos Institute is therefore partly correct when he argues that 'in the Indian context, some things have no relevance'. These 'things' are workers' exposure to asbestos, fulfilling legislative requirements, monitoring of the environment and industries' commitment to safe production techniques. These are all symbolic performances in which it is the 'appearance of doing' that matters rather than a commitment to environmental or occupational health. This 'appearance of doing' is ultimately about facilitating economic growth at all costs and is, in effect, a deliberate flouting of regulation at state level which works in conjunction with the national government's determination to define asbestos as safe, and to limit the recognition of ARDs. This is part of an unstated agreement between industry and the national government which hinges on the assumption that economic growth is critical. As Mr Patel, retired Justice from the High Court of Gujarat explained:

> *The argument from the government usually is development and they use the word sustainable development, but the emphasis is on development. Their usual argument is that if there is development, then there will be employment, production and generation of the benefits of development. And the government will for some time condone the breaches. That is how things are happening. They say that we don't have the option to develop or not to develop. For example, if the* Blue Lady *had not come to India it would have gone to China and many other ships would be diverted to China. Shipowners would think that China is the place where there would not be difficulties and if you were to compete with China, we would have to compete on all aspects … So we have to make sacrifices for development. The argument boils down to: do we want to develop or not develop? If you go by all these [occupational health] standards, you can't develop.*

The alliance between the Indian government and corporate business and their ability to reach a compromise – at the expense of the workers – also influences the style of mobilization adopted by activists. In the case of BANI, this led activists to centre their activities around international protests and to overlook the potential for grassroots organizing. As demonstrated in the case of Shree Digvijay Cement Company, not all Indian anti-asbestos activism is conducted in this style. In 1997, the workers from Shree Digvijay Cement Company came to Paryavaran Mitra, an NGO that facilitates villagers' participation in Environmental Public Hearings (EPHs),[29] complaining about asbestos cement roofs and associated breathing problems. Paryavaran Mitra then sought to facilitate the workers by creating conditions conducive for the judiciary to file its own case or '*suo-moto*'.[30] It did this, not by becoming engaged as a political actor, but by guiding the workers in media presentations and showing them how to write a letter to the High Court. Echoing BANI's style of activism, this allowed the NGO to be involved in the mobilization, but not to have its name associated with it. The crucial difference between BANI and this NGO is, however, that Paryavaran Mitra is working, not with international activists whose interests wax and wane according to diverse political cycles, but with local people and workers. Paryavaran Mitra is focused on developing grassroots movements through which people can take control over their own lives. Its main aim is thus the transfer of technical scientific knowledge to people and, through creating greater awareness, the implementation of environmental reform. Paryavaran Mitra thus helps villagers and people wishing to protest by offering explanations of technical scientific material coupled with guidance on mobilization strategies and procedures.

On 8 October 1997, Anilkumar Powar – having been informally advised and assisted by Paryavaran Mitra – sent a letter to the Gujarat Pollution Control Board[31] stating that the manufacture of asbestos products by Shree Digvijay Cement Company was causing serious health hazards to people working and residing in the vicinity. Powar argued that there were visible asbestos and cement particles in the air as well as in the drinking water provided by Shree Digvijay Cement to its residential colony. In his affidavit, he also asserted that waste asbestos was dumped on factory grounds without adequate precautions, land was being denuded, workers were not provided with any information after their company medical examinations and finally, as a result of all these factors, that people residing in the vicinity of the factory were prone to ARDs. Indeed, Powar's own father had worked for this company and had died of lung cancer in August 1996. Other workers and residents of Ranip supported Powar's claims and wrote accompanying letters requesting immediate remedial action.

Shree Digvijay Cement declared that Powar's allegations were 'completely baseless', stemming from a sense of 'spite and vengeance' because Shree Digvijay had dismissed him for 'gross and serious irregularities'. The affidavit-in-reply states: 'In our factory, we are maintaining the safe dust exposure limits as prescribed under the Factories Act, 1948. Hence there is no chance at all for the persons residing in the vicinity/colony to contact [sic] diseases like cancer or TB due to the exposure of asbestos/cement dust.' It also denied that the surrounding areas had been decimated of plants, that asbestos particles were present in the water and air while specifying that, according to the WHO, the ingestion of asbestos particles was not dangerous. Shree Digvijay Cement also argued that the wet manufacturing process was completely automated and foolproof, with no fibres becoming airborne. Similarly, once the asbestos has bonded with cement, there is no possibility of exposure to asbestos fibres. Shree Digvijay's annual medical examinations, coupled with the fact that workers undergo chest X-rays every three years and its compliance with the Factories Act, mean that 'the view that [the] industry of the answering respondent can cause air pollution and diseases like lung cancer is absolutely theoretical and speculative'.

As can be seen, workers' knowledge of ARDs is pitted here against 'safe' production procedures and regulation. It is one set of understandings about asbestos positioned against another set of understandings about the same product. Clearly, someone had to make a decision about whose views held the most weight. Perhaps not surprisingly, in a case such as this, the answer was to be sought in science. The judge had the discretion to invite someone to be '*Amicus Curie*' or 'Friend of the Court' and to participate in the case, and, coincidentally Paryavaran Mitra was invited to perform this role. As *Amicus Curie,* the NGO 'requested hiring a national institute to prepare a report'. Despite the fact that Shree Digvijay Cement reacted strongly against such an external assessment, the NIOH was contracted by the Gujarat High Court to assess the health hazards. It conducted air samples at three sites around the cement factory and concluded: 'Fibre concentrations in the vicinity of the factory were very low and adverse health effects i.e. asbestosis, lung cancer and mesothelioma of pleura and peritoneum have not been confirmed at these levels' (NIOH, 1997, p2).

Paryavaran Mitra immediately contested this report as it was seen to 'favour industry'. It argued that the report was neither 'reliable nor scientific' because it did not include factors such as wind direction, and other micro-meteorological factors (Pancholi, 1997). Although the NIOH had taken samples over a 24-hour period, it had done this 6–9m above the ground and not, as Paryavaran Mitra pointed out, at breathing level. It had also not investigated water and soil contamination and had overlooked the production process, ignoring cutting and grinding activities. The High Court was, however, happy to accept the NIOH conclusions stating: 'We, therefore, prefer to place reliance upon the report of the NIOH.'

The NIOH report and the High Court decisions replicate other patterns of regulation and control of asbestos pollution: first, a symbolic exercise was carried out, air particles were monitored. Secondly, Shree Digvijay Cement was accorded the responsibility of regular monitoring. Both these two activities allowed the factory to continue operating as normal. The High Court ultimately decided that the grievance voiced by Anilkumar Powar 'does not appear to be completely acceptable'. It added, however, that the Gujarat Pollution Control Board should carry out quarterly inspections of the premises and, if necessary, instruct Shree

Digvijay Cement on any remedial measures (order dated 20 April 1999 for SCA/8617/1997 special civil application no 8617 of 1997, Suo Motu versus Gurarat Pollution Control Board). As is clear from this and other examples discussed above, the primary focus is on a symbolic process of monitoring, with no attempt by the state to question or challenge any of the assumptions which frame the manner in which monitoring is carried out. Instead, monitoring is seen as an end in itself.

Representatives from Paryavaran Mitra comment that 'it's difficult to work in this field'. As an activist-based organization, it comes under constant pressure from other NGOs, from government and from industry to focus on service delivery. Indeed, in contrast to BANI in Delhi, members of Paryavaran Mitra have not received any direct threats on their lives, families or work, but their position is not helped by having links with foreign organizations. Whereas BANI activists were protected in Delhi and other large cities through their international links, in Ahmedabad, international connections are seen as a disadvantage to activists. NGOs which survive through foreign financial support are labelled as 'not helping', as 'misleading the people' and as 'obstacles of development'. Because it is afraid that it will be labelled as 'anti-government' or 'anti-development', Paryavaran Mitra does not engage directly with BANI and is not a registered member of BANI. It does, nevertheless, have networks with other organizations that support BANI and relays any information it may have for BANI to use.

Raghunath Manvar, who runs an Ahmedabad NGO, also seeks to mobilize workers rather than follow international activist-orientated trends. He created the Occupational Health and Safety Association as a response to his personal experiences as a worker in Ahmedabad's thermal power plant. When he realized that his colleagues were dying 'day by day', he set about trying to understand what the problem was. He met with the NIOH director who explained the dangers of asbestos fibres. Armed with this knowledge, Manvar tried to get his fellow workers properly diagnosed, but found that doctors were not willing to explore the possibilities of occupational diseases. Instead, when undergoing medical examinations, workers were X-rayed, questioned about their smoking and drinking habits, and then diagnosed with tuberculosis. In order to counter the superficial and purely bureaucratic nature of these monitoring processes, the Occupational Health and Safety Association has undertaken medical training for workers. It shows them how to look for clubbing, how medical check-ups should be carried out for occupational diseases, emphasizing the importance of work histories. It has used this information to file a number of cases in the Supreme Court and in Gujarat High Court. The Occupational Health and Safety Association claims that, to date, it has been able to prove about 85 cases of ARDs. This is, however, a pyrrhic victory as only two of these identified cases have officially been recognized by the state of Gujarat and received state compensation. The Occupational Health and Safety Association operates on a shoestring (with the office located on Manvar's veranda) and has enormous difficulties accessing funding. This is, in part, because of Manvar's difficulty in dealing with international actors in English and, in part, because the Association is not officially registered as an NGO. Despite submitting an application in 2005, he had had no response from the government more than 18 months later, in 2007. This delay echoes the complicated procedural processes associated with claiming compensation as a result of occupational exposure. Both provide an indication of the lack of government commitment to occupational health issues.

Raghunath Manvar has developed his anti-asbestos stance, like Paryavaran Mitra, in relation to local workers' health, because of their exposure to local political and economic issues and through their own experiences of what is happening environmentally in Ahmedabad. They have therefore focused their anti-asbestos campaigns on mobilizing villagers and workers affected by asbestos. This is in stark contrast to the activists working in New Delhi. It also means that these activists have access to very different resources and networks which include villagers, workers, local authorities, the local media (which is more accessible than in Dehli), lawyers and researchers. In this regard, their strategy is very different from that of BANI. They are, as Long Martello and Jasanoff (2004) would argue, being driven by the 'local', by emotional commitments to people, places and livelihoods. It is their 'situated knowledge' that drives Raghunath Manvar and Paryavaran Mitra to challenge the Gujarat state and businesses on the health-related aspects of asbestos. Nonetheless, as is also evident in the above discussion, the discursive style and the terms in which these debates are carried out are based on scientific knowledge. In contrast with the South African example discussed in the previous chapter, local Indian government forums for addressing occupational health issues do not allow people's situated knowledge to be ranked alongside scientific knowledge.

Grassroots mobilization does not, however, occur in isolation and there are some links between the different kinds of campaigning. Paryavaran Mitra has informal relations with members of BANI's network, with information being fed both ways. Raghunath Manvar has been involved in some of BANI's activities, including attending meetings in Delhi and the high-profile, IBAS-organized, Global Asbestos Congress held in Japan in 2004. These different strategies have not, however, resulted in different successes. As indicated in this chapter, the Indian government's determination to frame asbestos debates in terms of safety and regulation and to define ARDs primarily in terms of asbestosis severely limits all possibilities of mobilization. In addition, processes of governance focusing on monitoring allow for a symbolic process which undermines people's health concerns and, coupled with complex bureaucratic procedural processes, continue to make it difficult for workers to claim. Caught between these two sets of negotiation is the NIOH, which seeks both to meet workers' needs and to fit within the government's desire to facilitate economic growth.

Conclusion

In seeking to attract foreign investment and international exchange, the Indian government has avoided regulating industries (Lipschutz, 2004) that have been able to bypass their social and environmental obligations, under the guise of 'monitoring' and 'regulation'. Given the absence of the government in the arena of occupational health and environmental pollution, NGOs such as BANI, Paryavaran Mitra and the Occupational Health and Safety Association have arisen (compare with Lipschutz, 2004). These have, however, been subject to control by the state through conventional and innovative means such as threats of western bias, bureaucratic slowness, institutional fragmentation, and by framing asbestos and risk as controllable processes.

The examples discussed in this chapter show how critical questions of identity are to the mobilization process. Drawing on Appadurai's notion that transnational civil society is simultaneously a project, a process and a space, the chapter has argued that anti-asbestos mobilization occurs across national, local, state, transnational and global spaces. Some Indian campaigners, particularly those based in Delhi, identify much more closely with transnational, global actors. For these actors, the process of building alliances, creating networks and struggle means that geographical, political and cultural differences between, say, New York or London and Delhi, are compressed. Other activists have identified more closely with workers and with local conditions of oppression. For these anti-asbestos activists, different networks and alliances come into play, often with an explicit rejection of the globalized anti-asbestos campaign. Forging ahead in more isolated contexts and seeking to influence local political processes, the geographical, political and cultural differences between Ahmadabad or Bharuch and New York or London are massive. Nonetheless, as Yuval-Davis (1999) has commented, it is important to realize that these activist identities are, despite apparent differences and fragmentation, co-constructed and all these actors are simultaneously both enormously distant from one aspect of mobilization while being intimately connected to another. The cosmopolitan and globalized nature of BANI's activism is, in part, possible because of the organization's remoteness from workers. Similarly, Raghunath Manvar's association with the workers is possible precisely because, as an uneducated former worker, he cannot engage with the international, globalized elite of transglobal activism. Ironically, this means that the co-construction of activist identities acts to expand the gulf between cosmopolitanized BANI activists and invisible sufferers of ARDs. It compresses some transnational geographical, political and cultural identities at the expense of local and national regional, ethnic, caste (and class) and educational differences.

Lipschutz argues that, despite desiring change from states, NGOs and other civil society organizations tend not to challenge the structures of the state. Their more conservative stances allow them to avoid charges of radicalism, socialism or terrorism, but means that their influence on the state is limited (2004, p231). This chapter shows the extreme difficulties NGOs experience when trying to challenge the structures of the state and when aiming to move beyond compromise and negotiation. The regulatory power of the state to define how asbestos is to be framed enables it to 'shut out' opposition. The state is, of course, not the only actor and it is increasingly pressurized by private companies which seek favourable climates for economic investment and by civil society and NGOs with a social justice framework (Keohane and Nye, 2000). Nonetheless, as indicated in this chapter, the work of BANI and other organizations seeking to end the use of asbestos is, in effect, too 'fragmented and diverse to wield significant structural power' (Lipschutz, 2004, p231). This fragmentation results, however, not only from the manner in which the state seeks to respond to social mobilization, but also from the ways in which activists and other agents of mobilization are situated within global 'spaces' and physical localities. While drawing on global connections works for activists in Delhi, it undermines similar work by NGOs in Ahmedabad. One central feature that makes activism and mobilization extremely difficult in both places is the absence of workers able to challenge either the state or the companies they work for (discussed in more detail in Chapter 6). Throughout the Indian asbestos campaign, workers are missing – as victims of ARDs, as grassroots organizers, as campaigners and as active agents on the shop-floor. While asbestos movements seek to shape workers as rights-claiming citizens, the state and industries continue to define them as inconsequential and invisible and, in so doing, undermine anti-asbestos movements' abilities to translate their vision and global processes into meaning and accountability for ordinary citizens. Instead, their responses remain localized and low-key. Ultimately, despite the co-construction of BANI and the Ahmedabad mobilizing strategies, the disconnect between BANI and the workers makes it difficult for activists to produce the 'Indian evidence' required to prove that asbestos is dangerous (discussed further in later chapters).

The past three chapters have all examined the diverse ways in which people are defined by national legal or political systems. They have explored the interactions which stem from these definitions and shown how ARD victims respond to, and challenge, different forms of governance. In these three cases, we see different manifestations of diseased identities. In Chapter 2, laggers' masculine identity and current lifestyles were undermined by the diagnosis of any ARD, regardless of how benign or serious doctors considered individual men's conditions. In the

South African example, explored in Chapter 3, asbestos pollution and disease form a platform from which to begin to engage in community participation, to reshape scientific knowledge and, in so doing, to strengthen both a local and a national identity. Here, ARD sufferers and other people exposed to asbestos demonstrate an identity shift from victim to an active participant in policy processes. In the Indian case discussed in this chapter, the identities are notable not by their active mobilized strategies, but rather in their absence. It is through processes of governance, informed by scientific, medical and legal understandings of knowledge, that ARD sufferers are denied any related form of identity. The following two chapters are explicitly comparative, exploring differences in mobilization strategies and emic experiences of disease and identity.

Notes

1   There must be consensus from all countries represented at the Rotterdam Convention before chrysotile asbestos can be included on the PIC list. As India, Pakistan, Canada and Russia provided substantial opposition, chrysotile has not been included (although amosite and crocidolite are) on the PIC list.

2   This willingness on the part of the Indian government to flout international opinion and to encourage the asbestos industry appears ongoing. News articles report on India's exposure of shipbreakers to asbestos, as ships such as the *Blue Lady* are permitted to remain at Alang for scrapping. This, as journalists have pointed out, violates the Basel Convention on the Control of Transboundary Movements of Hazardous Wastes and Their Disposal.

3   ARDs, especially mesothelioma, are on the rise in industrialized countries and are expected to peak in the following decade. Extrapolating figures from Finland, the International Labour Organization (ILO) estimates that each year 100,000–140,000 workers die from ARDs. Conservative estimates, based on an amalgamation of studies and extrapolating results globally, estimate that the total deaths from the asbestos cancer epidemic will range from 5 million to 10 million (LaDou, 2004).

4   It comprises 28 states and 7 federally governed Union Territories. These subnational administrative divisions are the National Capital Territory of Delhi; the Andaman and Nicobar Islands; Chandigarh; Dadra and Nagar Haveli; Daman and Diu; Lakshadweep; and Puducherry. Geographically, India covers 3.29 million square kilometres, which means that spatially it is the world's seventh largest country. As more than 1 billion people live in India, it has the second largest population in the world.

5   Including the Factories Act of 1948; the Dock Workers (Safety, Health and Welfare) Act, 1986; the Indian Boilers Act, 1923; the Dangerous Machines (Regulation) Act, 1983; and the Environment (Protection) Act, 1986. Also, the Manufacture, Storage and Import of Hazardous Chemicals Rules, 1989, and other rules framed under the Environment (Protection) Act.

6   This scheme is limited to workers in receipt of wages that do not exceed 3000 rupees per month. Thus, someone who is diagnosed after being laid off, or was once, but is no longer employed, cannot claim. In the case of ARDs, which have a long delay between exposure and the onset of the disease, this works to the employers' advantage.

7   The Rotterdam Convention is a multilateral agreement between country signatories that promotes shared responsibilities for importing hazardous chemicals. It became legally binding to its 73 country signatories in 2004.

8   In 2004, when the 11th session of the Intergovernmental Negotiating Committee (INC-11) met to consider the inclusion of chrysotile asbestos on the PIC list, it was unable to reach consensus, as India, Russia, Zimbabwe, Columbia, China and others opposed the inclusion (Kazan-Allen, 2006; Joshi et al, 2006; Krishna, 2007a). The same occurred in 2008 (Ruff, 2008).

9   The World Health Organization (WHO), the International Labour Organization (ILO), the International Agency for Research on Cancer (IARC), the International Programme on Chemical Safety (IPCS) and the World Trade Organization as well as medical associations such as the Collegium Ramazzini.

10   Chrysotile is composed predominantly of magnesium while crocidolite and amosite have high concentrations of sodium and iron.

11   This research is based on the understanding that previous studies exposed animals to very high concentrations of chrysotile, resulting in lung overload. Bernstein and Hoskins (2006) also point to evidence that heavy and prolonged exposure to chrysotile does produce lung cancer.

12   Less prominent in the global debates are attempts to assess the significance of fibre mass versus the number of fibres.

13   It does not, however, define what this threshold might be. In practice, threshold levels vary according to government regulation and this works to the asbestos industry's advantage. For example, asbestos standards in Ontario are set at $0.1f/cm^3$ (fibres per cubic centimetre) which are said to carry a 'lifetime risk of 5 excess lung cancers per 1000 workers' and a worker's risk of two in 1000 workers contracting ARDs (Brophy, 2006, p17) whereas the legal threshold in India was $1f/cm^3$ (DGFASLI, 2005) and has only recently been brought down to $0.1f/cm^3$ (Ansari et al, 2007).

14   The Collegium Ramazzini is an international society of academics concerned with critical occupational and environmental medicine. It is committed to the promotion of health and prevention of disease. Named after the Italian physician, Bernardino Ramazzini (1633–1714), it is financially independent and has a membership of 180 elected physicians and scientists from 30 countries (Landrigan and Soffritti, 2005).

15   These include sites such as the White Lung Association, based in Baltimore (www.whitelung.org); the International Ban Asbestos Secretariat (http://ibasecretariat.org); Mining Watch (www.miningwatch.ca); the Mines and Communities website (www.minesandcommunities.org) and so forth. Juxtaposed to these websites and articles are the web pages created by, or supported by, the asbestos mining

companies. The Canadian chrysotile mining industry is perhaps the strongest of these organizations and it supports a wide range of activities aimed at generating confidence in asbestos and asbestos-related products. The Chrysotile Institute is dedicated to promoting the safe use of chrysotile, to disseminating information and advice to chrysotile producers and users, and to inform and advise the general public, the legal structures, media and other concerned peoples. In addition to hosting its own website (www.chrysotile.com), this organization creates glossy newsletters which relay scientific conclusions that reflect positively on the use of chrysotile and that engage directly in policy assertions and politics.

16 The preventative and protective measures for the safe use of asbestos have been laid out in Convention 162 since 1986. However, in 2006, the ILO adopted a new resolution regarding the use of asbestos which implicitly suggests that the ILO favours the banning of asbestos (Chrysotile Institute, 2007).

17 The IPCS is funded by the WHO, the ILO and the United Nations Environment Programme (UNEP), and is housed at the WHO headquarters in Geneva. The IPCS aims to establish a scientific basis for the safe use of chemicals while also supporting and strengthening national capacity.

18 This organization has strong links to the Canadian Asbestos Institute.

19 The history of BANI has been well-documented and therefore is not detailed here (Castleman, 2002; Kazan-Allen, 2003b).

20 The Times News Network, India, pointed out that the Indian asbestos industry was co-funding this research with the Indian government. In addition to being able to influence the research design, it will have privileged access to the reports prior to publication (Sethi, 2007).

21 The 'healthy worker effect' refers to companies' tendencies to keep only healthy workers on as full-time employees, while using part-time and casual workers to do the hazardous jobs; and workers' tendency to voluntarily stop coming to work once they get sick as they know that they cannot perform satisfactorily. This relieves management from having to record workers' illnesses and from having to retire them officially. The overall effect is that there is no company record of workers contracting illness during their period of employment.

22 Its members include Kalyaneswari (Kolkata), Toxics Link (Delhi), Paryavaran Suraksha Samiti (Gujarat), Mines, Minerals and People (Delhi), Mine Labour Protection Campaign (Rajasthan), Banjara Development Society, Greenpeace (Delhi), People's Training Resource Centre (Gujarat) and Occupational Health and Safety Association (Gujarat).

23 Originally known as the SS *France*, and then the SS *Norway*, this ship was turned away from Bangladesh in 2006 because it was considered too toxic to be dismantled there (see Ganguly, 2007, for a full account of the ship's movements in search of a place to be dismantled without undergoing prior decontamination).

24 As an international treaty, the Basel Convention on the Control of Transboundary Movements of Hazardous Wastes and Their Disposal aims to ensure that developed countries do not dump hazardous waste in less developed countries. It seeks to ensure that such materials are managed in environmentally sound ways without

being transported long distances from their origins. Ships going from one national jurisdiction to another are subject to the Basel Convention.

25   This is the world's largest shipbreaking yard where workers are not covered by any legislation; they tend to be primarily migrants and are not provided with any protective equipment or training before dismantling ships. Environmental activists estimate that one in six workers at Alang suffer from asbestosis (BBC News, 2006)

26   Damage to the hull of the ship limited its seaworthiness.

27   At least one Gujarat doctor does not, however, accept these figures. He comments that he 'cannot expose the government', but that he has come across more than ten cases of asbestosis in the past few years.

28   Gujarat Factories Rules, 1963, Schedule XVII specifies that the number of workers exposed to asbestos should be kept to a minimum, that the area of activity should be clearly demarcated and indicated by warning signs restricting unauthorized access, the need for exhaust ventilation in any room where asbestos production takes place, the use of protective clothing and breathing apparatus, the regular testing of ventilation equipment, separate accommodation for personal clothing, washing facilities, a prohibition against the employment of young people and smoking, and regular air monitoring – once every shift and entered in a special register (A-252–A-258).

29   In 1997, the government of India made it compulsory for newly established industries to hold an EPH before getting environmental clearance from the government (under the Environment Protection Act, 1986).

30   *Suo-moto* is a legal provision that allows the judiciary to pick up on any issue reported in the media and file a case that can later become a Public Interest Litigation (PIL).

31   This organization provides environmental clearance before factories are established and monitors them once in operation. The procedures are, in principle, ways in which community and worker participation are encouraged and can be seen as spaces in which democracy is enacted.

Chapter 5

'Through No Fault of Our Own': Asbestos Diseases in South Africa and the UK

The doctor says I have the dust, but it's still small and must still grow ... the doctor said there's nothing wrong with me, but I can feel that there's a problem in my body. (Mieta Willemse, Prieska, South Africa)

The pain is very strong, it [is] right through my chest and between my shoulders. I feel like an empty vessel which does not have lungs and a heart inside. (Lenora Lands, Griquatown, South Africa)

I asked at the clinic, do people not get treatment for mynstof *[mine dust]? They said, we give pain tablets. There is no special treatment that they give for the dust.* (Katjie Smit, Prieska, South Africa)

It's a cancer, you don't recover ... we live on the pills. (Paul Wilson, Griquatown, South Africa)

It's so inevitable, you start with five brothers, then three [remain alive], then two, you just live with it, mate. You know [that you're going to die]. (Timmy Fortune, Barking, UK)

We are so used [to the diagnosis], we don't take it seriously, rather we take it like ordinary people, healthy people. (Rik Matlu, Prieska, South Africa)

Once you've got the disease, it's never going to get better, it's just going to get worse. (Gavin Knowle, Barking, UK)

When they find out they've got the disease, it tears the soul out of most of them. (Jimmy Croft, Dagenham, UK)

Introduction

This chapter provides a comparative exploration of people's experience of asbestos-related diseases (ARDs) and how this impacts on their gendered identities. As demonstrated in the previous chapter, which focuses on India, workers are often not aware that they are being exposed to asbestos. This results in a situation where many people experience the symptoms of ARDs without formal medical recognition of their condition. This chapter is concerned, however, with massive exposure to asbestos in South Africa and the UK in situations where the dangers of asbestos have become widely known long after people's exposure. It focuses on people who came into direct contact with asbestos; working as thermal insulation engineers (or laggers) installing asbestos insulation and as asbestos miners or within asbestos processing plants. Many of these workers were employed by Cape plc, a company which dealt with asbestos for about 100 years, mining in South Africa and manufacturing and processing in the UK.[1] Its activities have left a wake of questions regarding environmental pollution, ARDs, compensation and treatment.

This chapter examines the legacy of these actions, exploring the lives and experiences of poor, proletarianized men and women in rural South Africa and of working-class, politically disempowered men and women in the UK who, despite the very different country contexts, were all exposed to asbestos during the middle of the 20th century. As is evident in the earlier chapters, ARDs have long concerned academics and there is a wealth of material on medical epidemiology (Selikoff et al, 1964, 1968; Selikoff and Hammond, 1975; Selikoff, 1977; Mossman and Gee, 1989), on litigation and corporate social responsibility (Newell, 2001; Coombs, 2002; Ward, 2002; Meeran, 2003), on historical processes detailing the power of mining companies (Myers, 1981; McCulloch, 2002) and on the social impacts relating to health, compensation and rehabilitation (Hessel and Sluis-Cremer, 1989; Felix et al, 1994; Randeree, 1998). The literature focusing on social impacts has stressed how people obtain compensation, the difficulties of accessing grants, the frustration of inadequate medical facilities, rehabilitation of the environment and safety procedures, legal processes and medical treatment for severe forms of ARDs (J. Roberts, 2002; Braun et al, 2003; Gravelsons et al, 2004; O'Regan et al, 2007). Much of this material relies on formal definitions of medical and legal categorizations of disease and examines severe cases, failing to investigate how victims themselves interpret and experience less dire forms of ARDs. This chapter addresses these absences in the literature, through an anthropological approach that focuses on meaning and subjective interpretations. Using emic perspectives of how people experience, interpret and respond to ARD symptoms, it investigates bodily experience, social relationships, kinship and gendered identity. As suggested by the quotations, this chapter provides a comparative perspective on how people in the UK and South Africa give meaning to their experience of disease. Focusing on collective responses extends the understanding of ARDs and illness beyond a biomedical analysis. The comparative perspective demonstrates how people's similar experiences are framed differently in the two countries through the application of different medical, legal and political instruments. The chapter thus argues that the political economy of ARDs is intimately connected to how different governments frame these diseases; in so doing, it demonstrates the significance of legal processes and compensation for people's identity and shows how illness, although not medically defined as debilitating, can profoundly affect – and be affected by – the social and gendered identity of people.

Company and country profiles

The UK and South Africa have a long and intertwined history, characterized by more than 100 years of British colonial occupation and racial segregation. South African independence occurred in 1960, although the election of the National Party in 1948 saw the implementation of apartheid rule. However, as McCulloch (2002) has shown, relationships between Britain and South Africa continued to facilitate the growth and expansion of UK companies. Cape plc was one such company, that extracted raw materials in South Africa – with scant regard for the safety of its workers – and that processed manufactured goods in London's industrialized zones, showing similar disregard for the resident population, until high levels of ARDs forced the closure of UK factories in the 1960s (McCulloch, 2002; Meeran, 2003; Gravelsons et al, 2004).

The Cape Asbestos Company Limited was incorporated in England in 1893 with the aim of mining, processing and selling asbestos-related products. It operated several asbestos mines in South Africa, including blue asbestos (or crocidolite) mines, with a mill in the Northern Cape and brown asbestos (or amosite) mines in the former Transvaal. In 1896, Cape opened its first factory in London processing 'yarn, cloth, millboard, steam packings, ropes and cordage' from asbestos (Competition Commission, 1973, p135). From 1899, Cape operated a number of factories in England for processing and manufacturing asbestos products. After an initial period of financial difficulty, Cape's operation was bolstered by the need for fireproofing materials during the two World Wars (Competition Commission, 1973). During these years, Cape produced insulation products, gas filters for respirators, fireproof lining boards intended for warships, specialized brake linings for armoured vehicles and fireproof clothing.

Demand for asbestos products remained high in the decades following World War II as expansion occurred throughout the UK. Although some products (such as asbestos yarn and fabric) became less desirable after the 1950s, asbestos insulation board was highly sought after and the market expanded considerably. Cape also extended its production of friction materials, particularly brake linings. As a result, the UK Competition Commission noted in 1973: 'Cape is the largest producer of amosite fibres and is responsible for over 90 per cent of total world production. About 30 per cent of Cape's output comes to the United Kingdom, mainly for use in fire insulation board' (1973, pp136–138).

In 1913, Cape established a factory in Barking, London, in the UK. The Barking fishing industry had declined during the latter half of the 19th century and new industries increasingly moved into the area because the Barking Creek and the River Roding offered good river transport. The Cape factory was situated on the banks of the Barking Creek, amid chemical plants, power stations and heavy industrial plants. These were later followed by oil refineries and hazardous waste storage facilities. In neighbouring Dagenham, industrial development provided employment, good salaries and highly cosmopolitan lifestyles. Male workers were drawn to the area, bringing their families to settle there and taking advantage of the new opportunities – all of which relied heavily on asbestos – between the 1930s and 1960s. As shown in Chapter 2, Barking and Dagenham remain heavily industrialized. Home to 170,000 people (Audit Commission, 2007), the area houses the now derelict Beckton Gas Works, three highly contaminated power stations, one of the largest sewage works in Europe and a large electrical switching station. It is the 'back end' of London, or 'the whipping boy of the A13' as one asbestos sufferer described it. Although the Cape factory has been dismantled and a housing estate now covers the area where Cape unloaded and processed asbestos, Cape's activities have left Barking with a heritage of asbestos pollution. Industrial pollution, low education, low life expectancy rates and widespread experience of disease (Barking and Dagenham Council, 2002) have led some to describe it as a 'mesothelioma blackspot' (Gravelsons et al, 2004).

Whereas Barking and Dagenham are urban, industrialized, residential areas in the eastern part of London, Cape's South African involvement – and subsequent environmental pollution in the form of asbestos waste – was located in rural areas. As described in Chapter 3, when Koegas and other Northern Cape mines were closed, people moved to Prieska and other surrounding rural towns where they survived on a combination of disability or old-age pensions, occasional employment and through extended family relations. Koegas is today a ghost town and an abandoned mine. It is also, however, a site around which former workers and their families have mobilized and initiated a land claim. Many residents of Griquatown and Prieska look forward to being able to return to what they recall as an idyllic rural lifestyle (albeit formerly supplemented by very low mine salaries, see McCulloch, 2002).[2] Their desire to do this takes account of the asbestos pollution. It is based on their desire to 'go home' and on their belief that – as they have already been exposed to massive doses of asbestos – the fibres are already lodged in their bodies and they are destined to contract ARDs in the future anyway.

Recognizing risk: Banning asbestos

In 1929, in the UK, Barking Council initiated its first inquiry into the dangers associated with asbestos. As described in Chapter 2, Members of Parliament (MPs) were informed of the dangers of ARDs in 1930 and the 1931 Asbestos Industry Regulations were implemented. These regulations sought, unsuccessfully, to determine 'safe' exposure levels for long-term asbestos workers. Thirty years later, pressure from the industry and the successful marketing of asbestos as a product which saved lives led to legislation increasing the legal limit of exposure and consequentially enhanced workers' contraction of ARDs. In 1968, standards were adjusted downwards, and this was later estimated to reduce risk of asbestosis to 10 per cent (London Hazards Centre, 1995). Official – and appropriate – recognition of the seriousness of asbestos exposure only came in the 1980s, when the UK prohibited the trade, application and supply of blue and brown asbestos and the subsequent banning of asbestos in 1985/1986 and 1999 (Gee and Greenberg, 2002; Kazan-Allen, 1999). These changes in legal exposure levels reflect the politicization of risk within the UK, the power of the asbestos companies and the uncertain nature of scientific research. As demonstrated in Chapter 2, these factors have combined to produce what, in retrospect, appears to be a certain arbitrariness in decision-making and policy on hazardous materials and occupational exposure.

In South Africa, research on the medical effects of exposure to asbestos was initially suppressed by the asbestos mining industry, whose actions were closely supported by senior members of the apartheid state – who supported the National Party Government's racial capitalism and sought to protect the foreign exchange generated by asbestos production in South Africa (McCulloch, 2002). Awareness of the political and economic dimensions began in 1979 when, faced with the closing of the asbestos mines and mill and the widescale loss of jobs in Prieska, its residents formed Concerned People Against Asbestos (CPAA), focusing on improving people's access to compensation and local experiences of hardship (discussed in Chapters 3 and 6). In the lead-up to the democratic elections of 1994 and the end of apartheid, a concerted effort to deal with asbestos-related problems emerged as South African civil society organizations and rural communities questioned the dangers of asbestos exposure (Felix, 1991; McCulloch, 2005). The CPAA broadened its campaign to include environmental and health issues (such as abandoned asbestos dumps and the rehabilitation of mines, the prevalence of ARDs, doctors' attitudes and accessibility and the presence of asbestos in ceilings and roofing of houses and schools) and now sought to negotiate with various government departments (such as the Departments of Environmental Affairs, Water Affairs, Health, Housing, Labour and Minerals and Energy) and with trade unions such as the National Union of Mineworkers (NUM). As demonstrated in Chapter 3, working with medical scientists to document the levels of asbestos exposure and disease helped the CPAA to see the international dimension of the problem, to make contact with international lawyers and to consider asbestos in terms of human rights and corporate social responsibility. The transition to a democratic government, the introduction of a rights-based constitution and human rights lawyers' visits to affected towns thus also increased people's awareness of the political underpinnings of asbestos disease (examined in more detail in Chapter 6).

Post-1994, the democratic South African government has supported activities for legal redress. It has also sought to facilitate environmental rehabilitation and to provide improved health facilities, although financial constraints have made these challenging undertakings. Today, South Africans have little doubt about the dangers of asbestos and all asbestos mining is banned, while the use of manufactured asbestos products was phased out between 2005 and 2008.[3] Within the Northern Cape, certain towns and rural areas are acknowledged to have been polluted by asbestos and people from a variety of backgrounds have come together to seek solutions. Prieska, in particular, has been prominently featured in the media. The close proximity of asbestos mines, dumps and the former milling of asbestos in the centre of the town have heightened the likelihood of people contracting ARDs, and Prieska residents are ten times more susceptible to ARDs than the national average (Kielkowski et al, 2000).[4] In conjunction with other polluted towns and citizens, the CPAA addressed questions of exposure, risk and reclamation and initiated an international campaign to secure corporate justice. These activities resulted in a transnational court case, with 7500 South African claimants taking Cape plc to court in the UK. In July 2003, Cape plc transferred £10.5 million to the human rights lawyers representing these claimants, in accordance with an out of court settlement that had occurred on 22 December 2001. This legal success story (see Coombs, 2002; Meeran, 2003; Ward, 2002; Waldman, 2007) has enhanced opportunities for compensation as other mining companies respond to the threat of litigation (Thompson's Solicitors, 2003; Morris, 2004).

Despite attempts by both the South African and the UK governments to control – through banning – the use of asbestos and to limit the number of people affected with ARDs, the problems persist. This is because of the widespread presence of asbestos – as a waste product from both mining and manufacturing – present in the environments, because people's prior exposure to asbestos makes them susceptible to contracting ARDs in subsequent years and because of the vast quantities of asbestos lodged in residential buildings, factories and other urban structures. In both South Africa and the UK, ARDs have continued to increase, although future predictions are not available for South Africa (Groenewald, 2005; Carnie, 2007). In the UK, predictions – widely believed to be conservative – estimate that ARDs will peak between 2011 and 2015 with 2000–2500 mesothelioma deaths per annum (HSE, 2006).

Recognizing risk: Emic understandings of danger

In both the Northern Cape towns and in Barking and Dagenham, it is hard to assess how much people knew of the dangers of asbestos at the time of exposure. On the one hand, one commonly hears people assert their ignorance. For example, Katjie Smit, a mother in Prieska, says she did not know of the dangers of asbestos and now she worries that she brought the illness into the house through collecting firewood in the vicinity of the mines. On the other hand, some people working with asbestos recognized some of the risks to which they were exposed. South African mine workers knew that they risked contracting *mynstof*. For example, Gert Bogoswe, who worked as a driver transporting asbestos, had heard that asbestos was dangerous. There was, however, little he could do about it. After he was taken for X-rays, his company simply said 'no, you're not sick'. It was only later, when Bogoswe consulted a private doctor, that his diagnosis of *mynstof* was confirmed. Similarly, Paul Myers, who grew up and worked on Koegas mine, described how he took his own precautions, using an *ou lappie* (literally, old cloth) over his nose and mouth, or a 'nosebag', to filter some of the dust. Nonetheless, people's attempts to protect themselves were limited and far removed from scientific work on ARDs. For example, the relationship between smoking, lung cancer and asbestos has been well-established in the international scientific literature since 1955 (Doll, 1955). The risks of contracting lung cancer are, for people who both smoke and work with asbestos, multiplicative (Hammond et al, 1979). In other words, a smoker who is exposed to asbestos has a far greater chance of contracting lung cancer, with some authors suggesting that the risk is ten times higher (Gravelsons et al, 2004). This academic knowledge was, however, not disseminated within South Africa and never communicated to workers, who were encouraged to believe, throughout the apartheid era and contrary to their own experiences, that asbestos was non-hazardous. Many of the Northern Cape residents continued to smoke, initially not associating this activity with the development of *mynstof* and with increased risk. As described above, their political mobilization around asbestos issues has, over the past ten years, increased this awareness and many people now recognize that smoking (and, they add, drinking alcohol) facilitates the absorption of asbestos into their bodies.[5]

Laggers and former asbestos workers based in Barking and Dagenham were similarly aware of the dangers, although they were never officially informed that they were being exposed to danger by their employers or contractors. Nor were they supplied with any safety equipment. As one resident explained, 'we thought asbestos wasn't as dangerous [as it is]. We thought one in 100 would get it. But others would not – we all have examples of men who worked with asbestos, who smoked, drank and were fine.' Nonetheless, many workers were implicitly aware that there were risks involved and took their own precautions. For instance, many laggers developed a constant dislike of dust and always sought to dampen down any dusty materials before they worked with them. One man described working with calcium silica boards that had to be cut and installed for insulation: 'I always used to dip them or get a hose to get rid of the dust. I've always had a bit of a thing about dust with everything so dusty. I would take a deep breath and dip it, then come back and breathe.' Others refused to enter very dusty places, drank bottles of milk in order to line their stomachs against the asbestos and wore handkerchiefs over their mouths. In contrast to the South African workers, very few of the laggers smoked, precisely because they were aware of the potential connections between smoking and lung cancer (Hammond et al, 1979).

In the research sites, in both countries, people developed their own categorizations of asbestos and risk in order to believe they were avoiding danger. These categorizations were based partly on their own experiences, and partly on the ways in which companies promoted asbestos. Thus, in the UK, people living in and around the Cape factory in Barking believed that blue asbestos (or crocidolite) was very dangerous, but that the white asbestos (or chrysotile) emanating from the factory was safe. The theory that blue asbestos was responsible for the most virulent forms of ARDs – namely lung cancer and mesothelioma – has been seen as a 'particularly successful line of defence' used by asbestos companies in the 1980s. Workers 'whose jobs depended on asbestos were, like the frightened public, glad to embrace the theory' (Pollitt, 1982, p6). By 2006, when my research was conducted, most Barking and Dagenham residents reflected, in disbelief, that they had accepted this explanation and had, as a result, not worried about the dust spewing out from the factory or about their working conditions.[6] In contrast, South Africans who were mining and cobbing blue asbestos in the Northern Cape were not exposed to this argument. As explained above, they understood *mynstof* to be a 'natural' disease. As early as the 1920s, residents of Prieska accepted that if someone contracted 'water on the lungs', they faced imminent death. These symptoms of mesothelioma were, however, undiagnosed and unrelated to the asbestos mines (McCulloch, 2005). It was not until the mines closed that people became aware of the extent of danger and – as they had already lost their livelihoods – they were thus more concerned with asbestos left exposed or used in the construction of buildings and homes in the town. As discussed in Chapter 3, they assessed risk in terms of fibre length (the longer fibres being more problematic) and texture (shiny fibres were considered dangerous), giving no indication that they knew about the presence of microscopic fibres. This emic emphasis on length and texture meant that residents were not concerned about the equally dangerous microscopic and invisible fibres which surrounded them. The use of different discourses in each country meant, in both cases, that people continued to work and live with the risks that asbestos created. They considered their conditions to be natural in that the connections between occupation, workplace organization, occupational health legislation and disease were not explicitly recognized or articulated.

Nonetheless, in both South Africa and the UK, people had notions of harm and sought to minimize risk by identifying dangerous contexts and avoiding exposure to asbestos. In many respects, there was nothing people could do about the risk and – at the time when this research was conducted – they felt that as they had already been exposed, there was little point in worrying about current or future exposure. Because of the latency period between exposure and diagnosis, many people believed that they already had an ARD, even if they had not been diagnosed. They felt, therefore, that there was little they could do except wait and hope that their good health would continue. In both South Africa and the UK, almost all those interviewed knew people – usually close family members – who had contracted or died of ARDs. Thus, one thing that people did spend a lot of time doing was discussing their conditions, assessing their health, visiting doctors and debating the degree of compensation they could claim. As discussed in the following section, this meant that the formal, legal categorization of ARDs – and their accompanying degrees of severity – was of critical importance.

Categorizing and diagnosing ARDs

As demonstrated in the preceding three chapters, ARDs are difficult to pin down, both medically and socially. Primary areas of disagreement concern medical categorization, diagnosis and the severity of mesothelioma, lung cancer, asbestosis, pleural thickening and pleural plaques. Different medical, legal and governmental understandings of these diseases have been used in South Africa and in the UK during different historical periods. This variability in conceptual framing of ARDs also shapes people's experience of illness, possibilities to claim compensation and mobilization strategies, and has, as discussed in more detail later in the chapter, significant impact on their gendered identities.

Chapter 2 examines the processes leading up to the October 2007 House of Lords' decision that UK sufferers will no longer be able to claim compensation for pleural plaques. The remaining diseases for which people can legally claim compensation in the UK are pleural thickening, asbestosis, lung cancer or mesothelioma.[7] In South Africa, people who suffer from ARDs are eligible to receive a state pension if they have more than 40 per cent damage to their lungs (or what is known as second-degree illness). This provides a small but steady monthly income of R820 (approximately £60) per month. In addition, former employees of the mines can receive compensation from the government-run Medical Bureau for Occupational Diseases (MBOD). The MBOD pays compensation for asbestosis, mesothelioma and lung cancer (Myers, 1981, p241) and distinguishes between 'first-degree' asbestosis or pleural plaques (in the past it used these terms interchangeably) and 'second-degree' damage (the presence of asbestos fibres plus additional scarring on the lungs, either from tuberculosis or smoking).

In both South Africa and the UK, apparent similarities and overlaps between the definitions of asbestosis and pleural plaques raise questions about medical definitions and legal categorization of compensation. Medically, the various forms of ARDs are considered to be different kinds of diseases, rather than as variations in degrees of severity. Pleural plaques are not seen as being related to other, more severe forms of ARDs such as lung cancer or mesothelioma. Pleural plaques are billed as benign and inert and, as a result, are said to be unlikely to lead to further damage of the lungs. People suffering from these 'benign' pleural disorders are expected not to experience pain but may have some breathlessness and discomfort (Mossman and Gee, 1989; Rudd, 2002). UK doctors have produced a convoluted argument in which pleural plaques are seen as evidence of exposure to asbestos and it is this exposure which leads to other ARDs. Although quoted in Chapter 2, it is worth repeating this argument here:

> *Pleural plaques are not thought to lead directly to any of the other benign varieties of asbestos-induced pleural disease, nor to pose any risk of malignant change leading to mesothelioma. Their presence may indicate, nevertheless, a cumulative level of asbestos exposure at which there is an increased risk of mesothelioma or other asbestos-related disorders. On average, in the absence of any other evidence about exposure, it is reasonable to assume that subjects with plaques will have had higher exposure to asbestos than subjects without plaques. The frequency of development of other complications of asbestos exposure in persons with plaques is not a function of the presence of the plaques, but of the asbestos exposure that caused plaques.* (Rudd, 2002, p344)

Asbestos sufferers do not, however, draw a distinction between pleural plaques and other forms of ARDs. They do not accept, as Rudd argues, that other asbestos-related complications are not related to the presence of plaques. Their experience of ARDs, and hence their emic interpretations, emphasize the interrelated nature of these diseases. In both South Africa and the UK, people I spoke to were very clear that they saw the disease as progressive and interrelated. The following represent a small selection of the comments I heard:

> It doesn't stand still. The mine dust grows, it constantly gets worse. It's a dust which spreads in your lungs. (Sarah Johannes, Griquatown, South Africa)

> Everyone begins with pleural plaque, it's verkalking [calcification or sedimentation, becoming like stone]… on the lungs. It goes to asbestosis, then to pleural asbestosis, and to cancer – and it ends here. It happens to everyone, it ends with mesothelioma and you die. (Flip Barends, Griquatown, South Africa)

> It's one sickness, it's the same for everyone, it's just the stages. Everyone ends up with cancer. (Maria Berends, Griquatown, South Africa)

> They diagnose us in the beginning with pleural plaque but it becomes asbestosis or mesothelioma either in the hospital or in the following weeks. (Jimmy Croft, Dagenham, UK)

> They say pleural plaques are not life threatening, but what it develops into is. (Ben Smythes, Barking, UK)

These quotes indicate that expert medical diagnoses may be more subjective than doctors admit. This is supported by Wikeley's suggestion that medical practitioners were exercising individual judgments when making their diagnoses of asbestos diseases, and that doctors use the term asbestosis to cover a range of asbestos conditions (2000, p116). This is contrary to much medical opinion, which represents the diagnosis of the various ARDs as clear-cut and unambiguous. Nonetheless, some medical personnel at the Griquatown hospital recognized the difficulties of diagnosing asbestos-related conditions, saying that 'as soon as you diagnose it [mynstof], it's basically at the end stage [mesothelioma and imminent death]'. Diagnosis is complicated by the delay between exposure and the onset of ARDs, by the extremely rapid deterioration experienced by mesothelioma sufferers and by the specialist skills – that are not always available in South Africa's rural areas – required to read X-rays. In addition, high rates of tuberculosis (TB), asthma and allergies are characteristic of the area, and made worse through smoking tobacco or hemp rolled in newspaper and breathing woodsmoke from cooking fires in the homes. This leads doctors to begin by testing for TB. Matters are further complicated by the links between TB and HIV/AIDS and the stigmas associated with these diseases. As there is no treatment for people with ARDs, particularly the less severe forms of disease, medical staff see little point in investing significant time and expense on identification of this disease. Indeed, patients themselves are often reluctant to travel away from home for testing and can ill-afford the minimal fees associated with ambulance transport and hospital costs (see Chapter 3). Thus, as a medical sister at the Griquatown hospital explained, 'if you've had TB once, then it stops with TB. Doctors and hospitals don't bother to look for further explanations; they don't explore life histories or work histories.'

Medical diagnoses are also intertwined – and in the process reified – by the compensation associated with particular forms of ARDs. The MBOD's approach, outlined above, has been to compensate 'first-degree' asbestosis (also called pleural plaques) and 'second-degree' infection (asbestos infection plus additional damage to the lungs, either from TB or smoking) as well as for lung cancer and mesothelioma (Myers, 1981). As a result, most workers expected – and received – compensation in the form of two bulk payments.[8] The first compensation payment was generally made after a person had been diagnosed with first-grade asbestosis, with a second payment if and when the disease progressed to mesothelioma or if the person contracted pulmonary tuberculosis. This MBOD categorization differed from the UK legal and medical system. When I interviewed them, senior medical specialists in both South Africa and in the UK commented on the laxity or 'generosity' of the MBOD criteria. One leading UK research physician specializing in lung cancer who examined 500 of the MBOD records, observed that people who had received ARD compensation had 'loads of other things' wrong with them. Another medical professor from the respiratory unit at a large Johannesburg hospital, who had worked in the Northern Cape on ARDs, explained that smoking and TB are the primary causes of pulmonary disease. He therefore believed it possible for people to 'smoke their way into receiving an MBOD payment'.[9]

UK asbestos sufferers also stress that their experiences and beliefs diverge from medical definitions of disease. Like South Africans, they blame a host of other symptoms on the presence of asbestos in their lungs. Bobby Jones said, for example: 'I disagree with Dr Dunne's diagnosis. He said I won't get asbestosis, it's not that bad. But why am I like this [coughing, breathless] now?' Similarly, laggers argue that the presence of pleural plaques on their lungs is an indication of their extensive exposure to asbestos. Chapter 2 details the laggers' insistence – contrary to medical opinion – that pleural plaques indicate the presence of other, worse, diseases dormant in their lungs. They support this argument by pointing out that men they knew and worked with are initially diagnosed with pleural plaques, but inevitably die of mesothelioma and lung cancer. These men also believe that doctors are misdiagnosing them, seeking deliberately to underestimate the significance of their disease: 'Pleural plaque is a trick, they diagnose us with pleural plaque to pay as little as possible.' Irving White, for example, was 'compensated for pleural plaque, but he has scarring of the lungs, which I see as asbestosis, he has asbestosis and mesothelioma', said his friend James Long. As was the case in South Africa, some UK doctors were simply not interested in exploring the possibility of ARDs. Anne Longmore, for example, told her doctor that she had worked with asbestos, but he 'didn't want to hear'. Gavin Knowle spoke of how doctors at East Ham Chest Clinic recognized that people had ARDs, but because they thought of pleural plaques as being symptomless and non-progressive, they failed to tell their patients. Alasdair Packard's account, briefly recounted at the beginning of this book, succinctly demonstrates UK doctors' easy dismissal of pleural plaques.

Other Barking and Dagenham residents were told by medical specialists that they were not suffering from anything other than pleural plaques and that they should have no symptoms from this. Often patients did not agree with medical diagnoses. For example, Gavin's friend went to the hospital with swollen fingernails which he interpreted as clubbing. He was told that it was not asbestos-related. Gavin disagreed. 'I know the signs, I know it's asbestos.' Sometimes, as demonstrated in the following case, patients were informed that their case was not serious, but informal opinions voiced by other medical staff persuaded them otherwise. Mildred Smythes explained how her husband, Max Smythes (who had worked for Cape for less than a year when he was 15 years old), had been to see Dr Dunne complaining about pains in his chest. On his second visit, arranged because Max could not accept Dr Dunne's conclusion that he was not suffering from an asbestos-related condition, a woman doctor examined the X-rays before they met with Dr Dunne. Mildred looked at the X-rays with this doctor and asked about some marks on the X-rays: 'What's that? It looks like when the snows first come, like little flakes.' The doctor explained: 'It is because of the asbestos.' Mildred queried this, saying: 'But Dr Dunne says he hasn't got asbestos.' The doctor replied confidently: 'I'm sorry, but that's asbestos.'

Caroline Collins, another Barking resident whose mother and aunt worked at the Cape factory, has been diagnosed with pleural plaques. She has difficulty breathing which, her doctor insists, stems from emphysema. Caroline feels differently and says: 'I disagree with my lawyer that it [pleural plaque] does not affect your breathing. I have scarred lungs, damaged lungs. It must affect me; I don't have the same normal breathing air capacity.' Peter Bill went to two doctors, one of whom claimed he did not have an asbestos-related condition whereas the other said that he did. Ian Bayley's experience illustrates the extent to which doctors can disagree and the frustration that people experience when faced with these disagreements. His local hospital first identified his problem as asbestos-related fibrosis. He was then referred to the London Chest Hospital where, according to his wife, he was diagnosed with Alivaritus. She understood this to be a dust-related disease not linked specifically to asbestos, although no such medical term exists. But, she insists, she did not pay much attention to this diagnosis because Ian 'knew' that he had asbestosis because he had worked as a lagger's assistant and his manager had been diagnosed with asbestosis. He wrote to the doctor at the London Chest Hospital and requested another appointment. Ian's wife recalled that, after seeing him for 'one minute', the doctor said, 'it's definitely Alivaritus' and that he 'hadn't worked with asbestos for long enough' to contract asbestosis. Ian went back to his local hospital which specified that his condition was 'definitely asbestos-related'. The contradictory information made him angry and very depressed. Ian and his wife simply did not believe the London Chest Hospital's verdict that he was not suffering from an ARD. As a result, Ian told his wife that after his death he wanted a post-mortem: 'I want it to be stated on my records that I had this disease.' After his death, Ian's wife requested an autopsy. She was paid compensation, which she believes confirmed that Ian had an ARD (she has never been able to face reading the results). She commented that receiving the compensation made her feel better as it was a formal act of 'acknowledging that he's got it', but also that, as explored in more detail below, spending the money after his death made her feel guilty.

There is no doubt that the bulk of medical attention is focused on the more extreme forms of ARDs. Here, doctors have little difficulty recognizing the stressful experience and emotional difficulty associated with diagnosis. Clayson (2006) points to the 'burden' of medical interventions for mesothelioma patients: the fact that people experience unpredictable pleural effusions,[10] emergency admissions, traumatic interventions, have to visit multiple hospitals and consultants, often end up seeing junior doctors, that the internet provides a confusing array of 'best' medicine and that sufferers often do not know enough about their own situations. She argues that mesothelioma patients are particularly vulnerable to suffering. They have to live with uncertainty and, contrary to other forms of disease, this is a downward path with few stable periods. Patients have no sense of mastery over mesothelioma. In addition, many people suffering from ARDs know all about the symptoms and experiences associated with it and expect the worst. However, to focus exclusively on mesothelioma and its associated emotional distress fails to acknowledge that all forms of ARDs are highly stressful. In both South Africa and the UK, people not diagnosed with mesothelioma also experienced very acute suffering and distress in association with the process of diagnosis. As described above, the failure of medical staff to diagnose ARDs, particularly in the less acute forms of disease, increased people's concern and stress. Lee-Anne Fortuine in Griquatown commented:

> Some nights I am so concerned, if I could only know what it is. I really thought that if they said it's asbestos, then I would get treatment. It makes me worry, I would like to know what causes the pain. They say it is not cancer of the chest ... you can feel it easily, it's under the chest and behind my shoulder, it's a lameness and then the pain moves forward again.

Ultimately, for many South Africans and UK victims, it was not only about the compensation money and the clinical process, it was fundamentally about the recognition and acknowledgement – from their doctors and society in general – that they had an ARD. Given this, it is not surprising that many people spoke about their pain and the lack of diagnosis as a burden: 'It's a ... tiredness and a thick, heavy pain that won't go away. It's like you have something inside you, like a stone that's lying behind your shoulders,' said Annette Diamant.

Treating and fighting ARDs

Medical facilities in the rural areas of South Africa are not nearly as well-resourced as those in the UK's East London.[11] Nonetheless, facilities, expertise and financial resources are not enough to help asbestos sufferers and, even in places such as Barking and Dagenham, medical treatment for all forms of ARDs is limited. In Barking and Dagenham, no specific medical intervention is used for people with pleural plaques and some doctors have been reported 'not to believe in pleural plaques'. People in Griquatown and Prieska are provided with vitamin pills (known as 'strength pills', 'water pills', 'salt pills' and so on) by the clinic, with the aim of building up their strength and general health. This absence of specific care leaves asbestos sufferers to find their own ways of treating the disease and of fighting ill-health. In the arid, semi-desert environment of the Northern Cape, South Africa, the dryness of the air is seen to aid asbestos sufferers. Basie Pieterse explains, for example, that ARDs develop more quickly in wet climates. 'If you contracted it here, you must stay here. If asbestos gets wet, it grows, it becomes longer and it doesn't break. It takes lots of moisture. Asbestos is a wetness which swells, grows and blocks your lungs.'

Many people from South Africa – and from the UK (described below) – made this association between asbestos, moisture and dryness. In their accounts of the disease, asbestos fibres lodged in their bodies were seen to have alien-like qualities and a sense of agency over which people themselves had no control.[12] They argued that the fibres, if dry, would form a cocoon and develop a hard outer layer. But because asbestos continually needed water, it would draw moisture from a person's body. These emic interpretations of illness and its agency stem from people's physical symptoms, particularly, an incessant cough and ever-present phlegm that constantly made them thirsty. People therefore sought to keep the asbestos within their bodies moist and they did this through drinking a combination of herbal remedies and western medicine. Gert Bogoswe, for example, commented that 'when it burns and dehydrates me, honey helps to keep it moist'. Most people used local plant remedies – such as *Wynruit* (Ruta Graveolens), *Wilde-als* (Artemisia Afra) and *Kankerbos* (Sutherlandia Fruitescens) – because the bitterness of these plants was believed to help the body's pain, cleanse the body and strengthen the blood. Almost everyone drank cough syrup of some kind or other because this 'kept the lungs moist' and because 'other medicine did not have this oiliness'. One particular brand, Scott's Emulsion, was said to be effective in 'building the lungs up, the lungs are too flat [weak or compressed by the plaques]'. In addition, people drank a combination of honey and Zam-Buk (a salve comprised of oil of eucalyptus, camphor, thyme and sassafras, which is not designed to be taken internally); cooking oil and vinegar; or fish oil and vinegar because 'the oil takes the burning away and makes it moist' while the vinegar, like the herbal remedies described above, provides the bitterness to relieve pain, cleanse and strengthen the body. Most people also used various forms of salve designed for muscular pain – such as Zam-Buk, Deep Heat or Tiger Balm – which they rubbed on their bodies in an endless and pointless pursuit of relief from their discomfort.

Residents of Barking and Dagenham echoed these emic concerns about the disease. They, like the people of Griquatown and Prieska, linked ideas of moisture and dryness with asbestos. The association between asbestos and moisture is evident in Malcolm Barker's description: 'Asbestos grows in your lungs like a reef, like a coral island … It turns like a stone when they take your lungs out.' People also wondered if asbestos mutated their genes and could be passed on to their dependents. This is partly because of how they conceptualized their bodies fighting the asbestos fibres lodged within them. Many residents argued that their physical bodies went through stages that affected their ability to resist disease. If someone was very young when first exposed to asbestos, he or she was more likely to contract an ARD because young bodies were considered to be more vulnerable and less able to resist asbestos fibres. If the person was older when exposed to asbestos, then disease would progress more slowly as his or her resistance was stronger. Understandings of bodily resistance to disease were, however, not only aligned within an age spectrum. Some laggers understood their bodies to go through periods of regeneration, similar to the way one's skin is constantly shed and replaced, based on a seven-year cycle. These regeneration processes were attempts by one's body to return to its more youthful condition. This process or regeneration was stressful because, in seeking to refresh and reinvigorate itself, the body also created new vulnerabilities associated with youthfulness and, in so doing, reduced its established resistance to ARDs. Indeed, as one lagger said: 'I worry every seven years.' In this emic understanding, individual bodies acquire an agency and ability to determine the extent to which ARDs progress, as opposed to the biomedical, clinical argument suggested above. Notions such as these explained why some men lived to 90 without showing any symptoms of asbestos while others died in their early 30s. The application of this emic model, in which older people were more able to fight the disease but still experienced cyclical moments of vulnerability, enabled individuals to explain why some family members or friends may have experienced ARDs while they themselves showed no symptoms, or vice versa.

South African ARD sufferers sought to fight the disease by staying healthy. They tried to eat more vegetables and men tried, despite their difficulty in breathing, to get regular exercise by going for walks. As Rik Matlu, diagnosed with pleural plaques, commented, 'the further you walk, the worse it [your breathlessness] gets, so you just walk more slowly'. This attempt to stay fit and healthy was echoed in Barking and Dagenham. Ben Smythes, who was diagnosed with pleural plaques, described how his doctor instructed him to 'keep the elasticity of me lungs' through exercise. 'He said I should use the stairs, push myself to the limit, jog. This is what I've done.' Indeed, Ben has taken the doctor's message so seriously that he hasn't had a holiday in 15 years. 'I'm frightened that if I stopped, if I didn't keep going, I wouldn't start again.' In addition, Ben has flu injections every year to ensure that he does not get sick and have to retire to bed. He does not do anything that might cause him additional stress. He has not moved house and he tries to be relaxed at all times. This is because he was warned by the doctor that being stressed may 'bring it on' and that, in order to keep the asbestos lodged inside him at bay, he should not change the rhythm of his body. 'It is very important, it was drummed into me that stress can change me condition. I have trained meself to have peace of mind.'

Despite these tremendous efforts, Ben is unable to win his battle with pleural plaques. No matter how much he tries to stay fit, to remain calm and relaxed, ultimately – and in contradiction with the medical model – he believes that he is going to succumb to other more severe forms of ARDs and die. This knowledge is derived from his personal monitoring of his body – despite trying so hard to stay fit and unstressed, at his last medical examination he noticed that his breathing capacity has declined – and from his association with other Barking and Dagenham laggers suffering from ARDs. Similarly, in South Africa, where ARDs affect extended families and communities, collective knowledge of what it means to have *mynstof* is widespread. In the Northern Cape – as well as other former asbestos mining areas in South Africa – 'Virtually every resident has a relative or friend who is sick or who has died from what the residents call "asbestos"' (Braun and Kisting, 2006, p6). The widespread damage done by asbestos mining and production has recently been confirmed in a study which found that 'asbestos pollution from an industrial source greatly increases mesothelioma risk. Furthermore, relative risks from occupational exposure were underestimated and were markedly increased when adjusted for residential distance' (Maule et al, 2007, p1067). Working in the area around Casale Monferrato in Italy where an asbestos factory had been active from 1907 to 1985, researchers demonstrated that environmental pollution from industrial sources has an effect of 'alarming magnitude' on surrounding residents who are exposed to a third of asbestos workers' risk. Even 10km away from the factory, the relative risk was 'still remarkably high' (Maule et al, 2007, p1069). As a result, ARDs are experienced as simultaneously personal and collective experiences. In the following section, I explore the collective nature of ARDs and the manner in which these diseases influence – and are influenced by – people's gendered identity.

Asbestos disease and identity

Thus far, there have been some suggestions that disease and identity are gendered (generally, men fight ARDs through fitness and exercise, while women worry about transferring the risk to the home and their families), but not a full exposition on this complex relationship. In general, researchers have explored a range of ways in which science and medicine interact with gendered personalities. They have focused on the relationship between the scientific industry and gender; on gender and technology (Kohlstedt and Longino, 1997; Lohan, 2000); on how social and moral values permeate into medical science and the social history of medicine (Warner, 1995), but they have not adequately explored the gendered identities that arise when men and women engage with medical science as it impinges directly on their bodies (with the exception of the literature that explores new reproductive technologies).[13] Anthropologists have generally overlooked the interrelationship between occupational health, gender and identity, although there is medical material on how disease affects men and women differently (Davies et al, 2004). In what follows, I examine how industrial diseases impact on men's and women's lives in Barking, Dagenham, Griquatown and Prieska. Drawing on the concept of 'effects made by gender' proposed by Henwood et al (2006), I demonstrate that, in both South Africa and the UK, women's identities in the context of ARDs are strongly located in the home and are reinforced through women's domestic roles. Although women's experiences of ARDs were obviously devastating, the disease reinforced women's location within the home and did not directly challenge their household roles as wives and mothers. In contrast, men's positive masculine identities were inherently tied up with work and therefore with exposure to asbestos, but men's masculinity was simultaneously threatened through the negative associations with illness (especially in the UK) and through ethnic conceptualizations of men's responsibilities (in South Africa).

The 'effects made by gender' approach was conceptualized by Henwood et al (2006) in an attempt to explain why men and women respond in stereotypically gendered ways – with men expressing lower levels of concern – to environmental and technological hazard surveying. They sought, however, to avoid essentialist or universal, normative explanations which overlook the 'parallel, historical, trajectory of greater fragmentation, multiplicity, and fluidities in men and women's social identities and subject positions' (Henwood et al, 2006, p5). Instead of seeing men's and women's understandings of technology and risk as 'naturally' associated with particular genders, they argue that gender itself acts on men's and women's identities in ways that propel them towards gender-normative positions. The 'effects made by gender' approach enables the researcher to identify contradictory processes in which gender both powerfully regulates men's and women's social and cultural experiences, and simultaneously fails to regulate these experiences. Thus they argue that the 'marking out and contestation of the meaning of gender is one of the effects made by gender' (2006, p5). From this perspective, women's articulation of moral, caring values in relation to environmental and technological hazards is related to the ways gender acts to exclude them from men's discourses of technology, science, engineering, control and power. And similarly, men feel unable to voice doubts about the inherently positive potentials of technology, science and engineering or to show their lack of enjoyment of scientific mastery for fear of appearing 'gender inauthentic'. Such an approach expands theoretically on the work of Moore (1993) who argues that discourses of gender draw sharp distinctions between men and women, but then offer a range of positions that correspond or reject these stark positions. Individuals can selectively position themselves differently within this spectrum at different contexts and times. Thus, as suggested by Henwood et al (2006, p23), 'different ways of knowing may lead people to construct risk problems in ways that are recognisably gendered'. This chapter applies the 'effects made by gender' approach to the understanding of ARDs and examines how gender, disease and identity intersect. It argues, in the following sections, that people interpret their own and other family members' experiences of ARDs in stereotypically gendered ways, focusing more on men's and women's ostensible roles within society than on their own complex, multifaceted and fluid identities.

People's experience of ARDs provides an appropriate context in which to examine the 'effects made by gender' because these are not diseases that individuals experience in isolation. Although the diseases are not transferred through bodily contact or sharing of body fluids, the microscopic presence of asbestos fibres in the air, on people's clothes and in the vicinity means that there can be widespread contamination from any one person's exposure. In both South Africa and the UK, people experienced environmental exposure, where their mothers or fathers had worked in the factories or mines and come home with asbestos on their clothing and from general environmental pollution produced by the factories and mines. As is evident above, this collective knowledge of ARDs and their effects shapes individuals' responses to diagnosis. Almost everyone in Griquatown and Prieska, and many people in Barking and Dagenham, have family members who have died from ARDs. They know what it does to individuals and to their families. Their accounts of their experiences often focus on how the disease has destroyed their families. Nonetheless, as demonstrated above, in both the Northern Cape and London, families – and particularly male members of families – benefited historically from the economic opportunities presented by asbestos. In South Africa, although apartheid policy imposed racial restrictions on families' co-residence, men working on the asbestos mines in South Africa were able to live with their nuclear families (McCulloch, 2002) as opposed to other South African mine workers who were housed in single-sex hostels for 11 months of each year. In addition, and seemingly despite the shocking conditions, the workers and their families residing on, or in the near vicinity of, asbestos mining sites came to be referred to as 'one family'. They visited each other, knew each other, despite sometimes being of different ethnic and racial classifications. Ida van Wyk commented nostalgically:

> *Koegas was one big family, we lived very close to one another, everyone supported each other. We had many advantages. We did not have to pay for water or for our houses, we received food hampers. We also shared toilets, we shared communal spaces and kept them clean together. It was a better living than farm people, the money was more, the living standards were better.*

Even once the mines closed, people moved in groups to the towns and continued to live in close vicinity to one another. They also maintained their connections with the former mine community through attending the funerals of Koegas residents and workers. Whenever someone from Koegas died, people came from the Cape, from Johannesburg and from as far afield as Namibia to be present at the burial.

The historical association with the mines and with strenuous, dangerous work remained a source of pride for men who still asserted these positions and status at workers' funerals. Working on the asbestos mines had been a way for them to earn an income, to assist their families and to assert their masculine identity. Although it was difficult, many of these male workers were able to earn enough to support their families through a combination of asbestos mining and other livelihood activities (such as selling wood, farm labour, sheep-shearing and building). Healthy men working on the mines were considered to be the primary income earners and they fulfilled their masculine role as providers and household heads (McCulloch, 2002). As a eulogist commented at Bankie Gouws' funeral, his voice ringing out with conviction and strength: 'We were miners!' There was no doubt that working on the mines was a powerful experience. It enabled the men to secure their roles as husbands and providers within their nuclear families and to maintain extended 'family' relations with the community resident on the mine. Men working on the mines emphasized their strength and independence and this pride continued after their diagnosis of ARDs. Johannes Mbeka was, for instance, described by his children as 'kwaai' (literally angry, strict and intimidating, but also strong and proud). He did not want assistance in completing his forms for asbestos compensation and he did not wish to discuss his compensation. As his daughter recalled after this death: 'He argued that he worked for his money, he did not want to beg for money, he did not loaf, he worked for his money. He was a proud old man and didn't want to struggle.' Because this pride is vested in a combination of work and strength, and because of the need to fight the disease through action, many men continued working as long as possible. For example, Struis Berends said: 'I can't work in the way I did in the past. Now I have to work very slowly. But if you do nothing, it gives the disease a chance to climb into you. I do something so that I can fight it a bit.'

Similarly, in Barking and Dagenham, the idea of 'the family' was broader than that conceptualized in a nuclear family and incorporated an extended, largely patrilineal, descent group. This family ideology has grown out of living in the East End of London and was specifically linked to the Cape asbestos factory (and other large firms such as Kitchings). These large corporations were believed to be 'like a family' – an image which the companies actively encouraged. This paternalism was, as Abercrombie and Hill point out, primarily 'an economic institution concerned with ... organizing a productive unit and regulating relationships between subordinates and the owners of the means of production' (1976, p413). In providing workers and their families with additional resources, paternalism created a work ethic in which workers felt personally and morally committed to their companies. In addition, as described in Chapter 2, because many employees were related to one another through kinship and marriage, the companies acted as hubs of social activity: they had good sports facilities, held social events and sports days, celebrated Christmas and provided parties and gifts for their employees and their families. Until the late 1960s and early 1970s, men in Barking and Dagenham sought employment with these companies which, as the primary employers in the area, maintained a paternalistic relationship with their workers.

There is no such paternalistic relationship between companies and employers in Barking and Dagenham today. Nonetheless, the ideology of an extended family has retained its significance for many segments of the local population. For example, many laggers living in Barking and Dagenham stem from the families that moved to the area in the 1920s and 1930s to take advantage of industrialization. Chapter 2 describes how lagging was seen as a profession that embodied masculine pride. Laggers were proud men, able to support their wives and families, enacting the role of a family patriarch that extended beyond the nuclear family, supporting young men who entered the lagging profession, and they were proud because of the physical strength and fitness maintained through lagging.

These men, like their fathers, saw their masculinity as solidly grounded in their experiences as Dagenham laggers and as family providers. They positioned themselves as workers, husbands and fathers, defining their primary role as that of breadwinner. It was through their ability to bring in money – rather than through alternative visions of masculinity that emphasized nurturing and intimacy (Henwood et al, 2006) – that they were able to support their families and sustain their masculinity. Thus, as Timmy Fortune commented: 'Our father made us a family unit, this has continued to his sons.' In both the UK and South Africa, the men's masculinity is shaped through their parenting children, their role in protecting dependants and their ability to provide for their families (compare with Collier (1998) and Gilmore (1990)).

Undermining masculinity

However, as is clear from the earlier sections of this chapter, working with asbestos, whether in South Africa or in the UK, always holds the risk of illness. ARDs have decimated families and individuals have suffered enormously, both from the diagnosis of ARDs and from a combination of non-diagnosis and unexplained symptoms. In both South Africa and the UK, men's identity was bolstered during the periods when they were working with asbestos and 'unaware' of the consequences. From the 1950s, when the oldest men interviewed began working, to the 1980s, when asbestos was banned in the UK and mines began to close in South Africa, men's identity was bolstered in similar ways – through the establishment of a masculine identity that emphasized work, danger and technology in conjunction with provision for the family and extended patriarchal relations within a community. From the early 1980s onwards, when men in South Africa experienced massive unemployment as a result of the closure of the mines, and when workers in both South Africa and the UK began to experience the physical sensations they associated with ARDs, their masculinity was fundamentally undermined. Although these experiences and time periods were broadly similar, the conditions of work and safety legislation varied greatly between the two countries. Consequently, the manner in which their identities were undermined contrasted radically. In the UK, men's sense of masculine identity was undermined by the diagnosis of disease and by the increasing economic marginalization of their profession (discussed in more detail below). In South Africa, questions of ethnic identity coupled with people's economic situation destabilized men's sense of masculinity.

In terms of ethnic identity, most Griquatown and Prieska men classified themselves as either 'coloured' or 'Griqua' or both, depending on the context.[14] In hierarchical South Africa, where race had mattered enormously for many years, these were identities that were associated with many negative qualities. 'Coloured identity' was a construction of the apartheid government (1948–1994) which included everyone who was neither obviously 'white' nor 'black'. It was thus a definition which categorized people through negatives: in terms of what they were not. The Griqua had been subsumed under the apartheid government's broader racial category of 'coloured', which included the diverse conglomerates that resulted from people's interactions during the colonial era and other categories of people, such as Malays, not easily identified as either black or white. Griqua identity today is thus complex and heterogeneous, combining many different traditions and overlain by, and entwined with, apartheid planning and racial discrimination (Waldman, 2006a). In Griquatown, ideas regarding Griqua identity, place and social status are entwined in the notions of *boorlings* and *inkommers*. *Inkommers* are newcomers to Griquatown (literally in-comers) and are coloured elites. *Boorlings*, or people born in Griquatown, are by contrast, 'nothing people'. Over the years their identity has been changed and manipulated. Apartheid governance saw them, quite simply, as coloured (President's Commission, 1983).[15] Their Griqua heritage was largely dismissed by the Griquatown Council and by the central government. In addition, the process of proletarianization had led to increasing impoverishment, with *inkommers* occupying positions of status and formal leadership. The *boorlings* were thus reduced to defining themselves in terms of place: where they were from (or origin). Former mine workers and their families who were resident in places such as Koegas have been absorbed into this hierarchy as *boorlings*. Their Griqua heritage, practice of Griqua traditions and origins in small Northern Cape towns, coupled with their impoverishment, makes their alignment with *boorlings* appropriate. This ethnic identification means, however, that there are few positive attributes (other than religion and work on the mines) around which men can assert their masculinity.

In Griqua cultural ideology, the categories of Griqua *boorling* and coloured *inkommer* assume primary importance and are gendered in ways which undermine men's masculine roles as household heads and providers for nuclear families. Although Griqua society is not structured along matrilocal lines, Griqua ethnic principles establish women as 'true' boorlings while men are considered inkommers to the nuclear family, but boorlings of the broader Griqua kin group. In ritual, men are positioned outside the home, because, whereas women are ideologically bound to the house, men are free, sexually unrestricted and expected to act accordingly. All husbands, regardless of their ethnic origins, are considered to be *inkommers* to women's families and, as such, cannot command much respect. Furthermore, as *inkommers*, not much is expected of these men. *Inkommers* cannot be relied upon to do what sons born to the house will do (Waldman, 2003).

These ethnic categorizations are not as strong in Prieska which offers people many other forms of identification (demonstrated in Chapter 3). Although Prieska is comprised primarily of people who identify themselves as Griqua and/or coloured (many of whom are related to people in Griquatown), the population of Prieska is more diverse, with greater Xhosa influence and the *boorling/inkommer* categories more diluted. While both Prieska and Griquatown suffer from unemployment and rely heavily on state pensions, Prieska offers more employment opportunities than Griquatown. As Prieska is located on the main railway line from Namibia to Cape Town, it is easier for people to travel to and from the town and many Prieska residents have attended university in Cape Town. Although Prieska residents can identify as educated nurses, teachers and government representatives (a category reserved for *inkommers* in Griquatown), as politicians active in the African National Congress and other political parties, as successful farmers, builders and entrepreneurs, former asbestos mine workers are excluded from these identities because of the lack of education and low working-class status. They are thus primarily unemployed. This means that, for former mine workers in both Griquatown and Prieska, there are few sources of positive identification available to them. Status, for them, remains linked to their dangerous work on the mines and, ironically, in their ability to bring in asbestos compensation and disability pensions. Gender relations, structural conditions of employment, and the manner in which ARDs are recognized and compensated thus combine to reinforce men's gendered identity as one that is bolstered through their association with technology and their powerful mastery over the physical world.

For many men who worked on the asbestos mines, there is – ironically – a degree of status associated with the receipt of *myngelde* (literally 'mine money'). Asbestosis sufferers are considered to be rich, albeit briefly, and hence valuable to their families. As mentioned above, former mine workers can access MBOD compensation, which provides two crucially significant and substantial payments. For example, during the late 1970s and early 1980s, asbestos compensation varied from R6000 to R30,000, depending on an individual's racial status, position held and employment duration (Myers, 1981, p241).[16] Once diagnosed with an ARD and awaiting MBOD compensation, men are entitled to various economic 'privileges' with shopkeepers advancing credit and encouraging them to open accounts (usually without advertising the interest charges and thus providing rather dubious 'benefits'). In addition, people classified with second-degree damage to their lungs can register for a state disability pension or *ongeskikheids* pay of R820 per month. This provides a reliable source of income which enables families to survive and which is relatively lucrative in the economically depressed environments of Prieska and Griquatown. Thus, although no longer active as miners and not working, the official government framing of pleural plaques as a form of disability allows these men to claim compensation and – in so doing – to continue to fulfil their masculine role as household providers.

Issues of identity and disease in the UK were similarly gendered, although ritual and ethnic identities were less significant. Here, because of men's work as laggers, it was their identities as masculine providers and, through this, their work relationships and associations with technology that were threatened. Although doctors advised that men with pleural plaques could continue to work, the way in which men responded to this information was not straightforward and did not accommodate this medical understanding of disease. As soon as a lagger was diagnosed with pleural plaques, he felt his ability to work and hence his identity to be challenged. As one Barking resident put it: 'When you get told you've got pleural plaque, you don't want to take money out of the family coffers, don't want to buy a pair of overalls', because of his belief that he will no longer be able to bring in an income. As suggested in this comment, so devastating was the news that it called for immediate changes in the men's lives. It was the diagnosis of pleural plaque – and hence the acknowledgement that you had 'the disease' – that caused distress and disruption. Prior to the diagnosis, and despite sensations of inexplicable pain, men fought the disease by refusing to acknowledge that they had it. Many men avoided – in so far as it was possible to do this – having regular medical check-ups[17] and avoided discussing their (inevitable) contraction of ARDs. By not recognizing the disease inside their bodies, they could pretend that things were still fine. As soon as they acknowledged the presence of asbestos in their lungs, they were – because of the way they interpreted the disease and because of their disagreement with medical models of ARDs – forced to consider their imminent deaths. They refer, for example, to Timmy Fortune's father who was told that he was fit to work as a lagger one year before he died; six months later doctors said he could do light work. His rapid decline and death illustrates what every lagger knows: once you have pleural plaques, you are going to die.

As Ben Smythes put it: 'He's got a cancer inside him that could explode at any minute.' Jimmy Croft explained in more detail: 'The disease progresses very quickly from pleural plaque to pleural thickening and asbestosis. You are incapacitated; you can't work, and have to live off the state. Because you can't work, you don't have any money.' The underlying idea being expressed here is that receiving assistance from the state undermines men's masculinity. Welfare systems in advanced industrialized countries such as the UK are, as Gordon has argued, inherently gendered and are created to reinforce male-breadwinner families by providing – only temporarily – for the collapse of these families and by encouraging men back into the workforce as quickly as possible. Assumptions about masculinity thus make it 'unthinkable for able male welfare recipients not to work' (1990, p11). The characterization of masculinity as breadwinning and independent is therefore reinforced through the welfare system. In addition, the contractual nature of the laggers' work means that they lack social protection mechanisms (discussed in Chapter 2). As sick benefits, injury compensation, and unemployment insurance are not secured, unemployed laggers have to rely on the welfare system. Thus, if they are not working, they are not earning. Men who had supported large families and played active community roles now found themselves dependent on others for even small luxuries like cigarettes.

For these reasons, laggers resident in Barking and Dagenham tended to initiate legal proceedings and to claim compensation as soon as they were diagnosed with a legally recognized disease. Although, as discussed in Chapter 2, they could request either provisional damages or go for a full and final claim, all the men interviewed favoured the full and final claim. They argued that: 'If you go for an interim payment, it doesn't help you when you have a future incapacitation. Therefore we go for a full and final settlement – if you get £20,000–30,000 you can invest it and use it when you have to stay at home and can't work.' In addition, if they waited until they were sick, it would be too late: 'You're in debt by then, living on handouts.' The logic behind this was that the men aimed to use the compensation money to establish themselves in a new line of business which would enable them to continue to provide for their wives and families. In practice, however, the payouts that men received were minimal. Payments issued in 2005, when it was still possible to claim for pleural plaque, were in the region of £3500–4000 for provisional damages and full and final settlements received £6000–7000. Clearly, this is insufficient to establish a new line of business. Given this, there was no possibility of maintaining their identity as men, as providers for the family or as proud patriarchs of extended families.

Once diagnosed, many men stopped working and withdrew from the trade union and fortnightly laggers' meetings. This meant that they lost touch with their lifelong friends and from 'family', separating themselves from their support structure. The men retreated from the company of other laggers because of the strong focus on ARDs at these meetings: discussion centred on laggers' diagnosis, claims and deaths. Although jovial and the laggers were full of ironic humour when discussing these issues, there was a strong recognition that they were all vulnerable and likely to be personally affected in the future. Laggers who had been diagnosed found themselves being constantly watched by the other laggers for symptoms. Meetings thus served as reminders of their own situation and as warnings of other asbestos sufferers dying. Disclosure of a diagnosis was thus closely related to the destruction of the social person, through a process of self-rupturing that pre-empted the disruption of the body.

In withdrawing from their support structures, they are also disengaging from the technology which is a central tenet of their positive identification. Faulkner (2000) suggests that engineers' intimate identification with work and its associated technology provides engineers with a 'separate reality' and, as demonstrated in Chapter 2, prevents assertions of other more positive identities when among non-engineers. Thus, on the one hand, men's positive association with technology, fitness and strength, is undermined by ARDs and this leads laggers to withdraw from their own social circles. On the other hand, the 'effects made by gender' inhibits the establishment of other masculine identities and the laggers are unable to articulate a masculinity that does not reinforce their association with lagging technology, and hence with disease.

ARDs and women's identities

Women in South Africa and the UK were in remarkably similar positions, in the sense that they articulated a stereotypical gendered identity in which women were located in the home and also that their involvement with asbestos has often been underplayed by themselves and others. Different ideological and economic justifications have, however, underlain these similarities. For South African Griqua women, their ethnic identity gave them a centrality within the home and ritual superiority over men. Nonetheless, many of these women worked on the mines, often in situations labelled as 'less dangerous' jobs. Although less overtly dangerous in the sense that women were not working underground, firing dynamite or doing hard manual labour, their role in cobbing asbestos fibre was inordinately dangerous for them and their families. As a result, many South African women suffered from more severe forms of ARDs, such as lung cancer and mesothelioma. Nonetheless, women in Griquatown and Prieska did not shape their identity according to this work in the public sphere. This was partly because they were seldom formally acknowledged as mine workers. Instead, their work was seen as supplementary to the main mining process and their employment was informal, generally linked to that of their husbands.

In addition, bringing money into the family was something they did out of necessity rather than something which was central to their role as women, as mothers and as Griqua. Because they had worked on the mines, they were – like their husbands and other male workers – able to claim compensation from the MBOD. This was part of their identity as mine people. As Elizabeth Pieterse from Prieska comments:

> *I was born on the mines, I grew up on the mines. I cobbed asbestos fibre, [moving] from mine to mine and stayed on the small mines. No, we are mine people, all of us. We worked on the mines because we were mine people. Grandmothers, uncles, aunties all lived there. We only lived on the mines.*

Unlike comments from male mine workers, Elizabeth's account – like that of many other women interviewed – was all about living on the mines and about the extended family members resident on the mines. Her identity was not located in the hard, dangerous work of miners nor was it centred on the ability to bring in money and provide for her family. Although other research in the Northern Cape shows that even married women aspire towards independence and can achieve a great sense of status when earning their own money (Waldman, 2006b), in relating their identity to ARDs women consistently focused on their roles as household keepers and as nurturers in society. This is in keeping with stereotypical gendered discourses that do not see women as engaging in heavy labour, dangerous work or with technological machinery. Despite the fact that many women did do these supposedly masculine activities while working and living on the mines, they articulate a positionality which is in keeping with the 'effects made by gender' approach of Henwood et al (2006).

In the UK most women's exposure to asbestos has been environmental (London Hazards Centre, 1995; Cancer Research UK, n.d.). A few women interviewed had worked at the Cape factory, involved primarily in clerical work and, as a result, not in direct contact with asbestos. This does not, however, mean that women were not susceptible to ARDs. Cancer Research UK estimated that, of the 2100 cases of mesothelioma diagnosed in 2003, about 400 were women. A study by Cancer Research UK and the UK Association of Cancer Registries (UKACR) examined cancer figures between 1995 and 2004 and found that women's incidences of mesothelioma had risen by 38 per cent during this time (cited in Thompson's Solicitors, 2007). Moore and Lenaghan (1995) argue that UK women's experience of ARDs has been overlooked by the medical and legal system as 'a combination of ignorance and prejudice in the medical, social and legal professions', and leads to inadequate investigation of women's exposure to asbestos. This constrains women from claiming compensation and denies them recognition of their condition. The London Hazards Centre points out that, underlying women's lack of diagnosis, is the erroneous assumption that cancer or asbestosis is caused by substantial exposure to asbestos (London Hazards Centre, 1995). For those women who have not worked at the factory, their risk of contracting ARDs stemmed from environmental exposure or indirect contact with the mineral via their husbands, fathers, sons, brothers, friends or boyfriends. It was through men bringing the dust home that these women were exposed to asbestos and contracted ARDs.[18]

In contrast to their husbands who argued that they had to put 'their house in order' (pay the mortgage, replace worn or damaged structures, purchase new furniture and so forth) so that their families would be 'provided for' after their deaths, UK women said that they had to 'redirect their lives'. Although some of the women interviewed were wage earners, none was the primary income earner. For the most part, women earned about half of what the men were earning. Although these women had complex identities that comprised, at least in part, their working lives and other individual experiences, they were also married to men who were strong household heads and who took control of the family. Thus, in relation to ARDs, these women articulated a primarily working-class identity locating themselves as housewives in their homes. After the demise of their husbands due to ARDs, these women spoke of their difficulty in re-establishing themselves. Having always had someone with them and providing for them, and someone they had to care for domestically, women found this new independence 'difficult to absorb'. Tracy Carole commented that she had to 'redirect her life' as she had been married for 40 years and had always had her husband with her.

Some women also expressed their guilt in relation to their husbands' exposure to asbestos and subsequent compensation. Melinda, Ian Bayley's wife, said that she felt 'guilty' about Ian's death. She felt that he had moved into Barking, become a lagger's mate and been exposed to asbestos because of her. Prior to meeting her, he had been a steward on a liner. 'It's my fault, if he hadn't met me he'd have gone back to sea.' Although, as discussed above, men were interested in compensation primarily to 'put their houses in order' and support their families, women who received compensation after their husbands' deaths felt ambivalence: on the one hand, the compensation provided a formal acknowledgement that the disease had really existed; on the other hand, women felt guilty about their independent spending of money that stemmed from their husbands' ill-health and subsequent deaths. 'The compensation made me feel better [it is a form of] acknowledging that he's got it [a form of ARD]. It would have been alright getting compensation while Ian was still alive. We could have done stuff with the money, made him more comfortable. Now the money makes you feel guilty, spending money because he died.'

Conclusion

Men and women are affected differently by ARDs, but both men's and women's experience of disease challenges the medical and legal discourse that categorizes disease according to perceived bodily experiences and places a financial value on this experience. This detailed examination of people's reactions to disease – including diseases medically categorized as benign – demonstrates deep anxieties about bodies, health and relationships. Ultimately, people's identities are intimately tied up with their ability to maintain relationships and to be socially connected. Contracting pleural plaques may not register on doctors' consciousness as debilitating, painful or distressing. Similarly, pleural plaques do not impress lawyers in terms of inhibiting people's working ability. But, as demonstrated in this chapter, the ramifications of this disease are felt socially and emotionally in ways not addressed in medical and legal discourse. The failure to recognize officially people's experience has diverse ramifications as people seek to rationalize and explain their bodily symptoms and emotional unease, to define risk and exposure in their own terms, to find their own sources of medication and devise ways they believe will keep them healthy and, ultimately, to continue to add meaning to their lives.

In both South Africa and the UK, women did not shape their identity according to work in the public sphere. Instead, their identity was vested in their household and primary caretaker roles. This meant, in effect, that women were in some ways shielded from the devastating impact of ARDs. Even when they were directly affected by the disease – through occupational exposure – their identity and sense of self was not directly undermined. For women in the Northern Cape of South Africa, supporting the extended families and facilitating ritual and ethnic roles – through work and symbolic ritual activity in the home – were very significant activities which enhanced their positive identity and status. For both UK and South African women, the identification of ARDs did not disrupt their roles or identities. Instead the additional roles of caring for sick relatives or taking care of their own diseased bodies fitted 'naturally' into their gendered roles. Although all these women had other composite identities, when interviewed in relation to their personal bodily experiences and familiarity of ARDs, they demonstrated conventional gendered alignments with the nurturing role of married women.

Men's inability to sustain a masculine identity associated with technology led, however, to a social withdrawal which preceded physical bodily deterioration. This social withdrawal was demonstrated bodily by men who had been diagnosed with ARDs. They were seen to change in personality, withdrawing into themselves. Referring to laggers diagnosed with ARDs in the UK, Jimmy

Croft said: 'They all get the same attitude, they jump at anything, they look out the window, they don't talk, they go into themselves.' Referring to how diagnosed men in South Africa react, Basie Pieterse said: 'It's as though he has a film over his eyes, as if he doesn't see. Nothing makes sense to him. He sees the world as though he is looking through a window.' The 'effects of gender' framework explains why men working on the asbestos mines in South Africa or as laggers in the UK located their sense of self firmly within their work, their strong physical strength and in being able to provide for their families, despite different legal frames and historical working conditions. For both these sets of men, working with asbestos was a positive form of identity, but it was one that was simultaneously threatened by the risks of ARDs. Although people in South Africa and the UK experienced similar physical symptoms, the use of a comparative perspective shows that the legal, political and medical framings of disease make an enormous difference to how people experience illness. It is these structural conditions which define the possibilities within which men and women can resist, claim compensation and seek recognition for their industrial work experiences.

The failure of the UK legal and medical institutions to recognize pleural plaques as an industrial disease is a social disaster for the men who worked with asbestos and who are now diagnosed with pleural plaques. The loss of work, together with being ill and the lack of alternative masculine identities, is a complex combination of forces which undermine their very being. In contrast, the more lenient categorization of pleural plaques as 'first-degree' asbestosis in South Africa provides former mine workers with an opportunity to access MBOD compensation and possibly, in time and with sufficient evidence of lung damage, to receive an income through a disability grant. Supplemented by the compensation received from Cape plc after transnational litigation and the possibility of future compensation as other mining companies respond to the threat of international litigation, *myngelde* (money received in compensation for ARDs) has significant status attached to it. Compared with the UK laggers who are unable to continue to provide for their families because of their 'disability', it is precisely because of their disability that former South African asbestos mine workers can continue to assert their masculine status and gendered role as family provider. The laggers have not, however, given up on their campaign to overturn UK legislation and to claim compensation for pleural plaques. The following chapter examines UK social mobilization in comparison with South African and Indian experiences. Using anti-asbestos mobilization as a lens, it explores the dynamics of mobilization, the politics of knowledge and the processes of governance of toxic protests.

Notes

1   Another source of exposure is the shipbuilding industry, about which much has been written in the UK and where asbestos mobilization is strong and ongoing.

2   McCulloch documents terrible living conditions in Koegas: rampant scurvy and tuberculosis, no mine housing and no provision for rubbish removal. In the 1950s, many workers became seriously ill – possibly with asbestosis – and were unable to work. As they had nowhere to go, they remained at Koegas. The widows of these men were relocated to Marydale in the mid-1960s and settled in a place which became known as the 'the lung location' and as a place of misery (2002, p114).

3   Trade Union Advisory Committee to the Organisation for Economic Co-operation and Development (OECD), 30 May 2005, (Draft) OHSE Sust/Dev Country Profiles for Asbestos, www.global-unions.org/pdf/ohsewpL_6.EN.pdf, accessed 16 June 2005.

4   For many reasons, including diagnostic difficulties, lack of adequate technology and difficulties tracing unrecorded migrant workers, it is very hard to estimate the extent of ARDs. Evidence of ARDs is characterized by 'gross under-reporting' and a lack of reliable data (J. Roberts, 2002). Nonetheless, Botha et al (1986) argue that in high exposure crocidolite areas (that include Prieska), the annual mesothelioma and asbestosis death rate for white South Africans was 542 per million as compared to 24 per million in the control area. Kielkowski et al (2000) report that the rates for ARDs in Prieska are 366 infections per million men and 172 per million for women, whereas national incidence rates for South Africa are 37 per million for men and 16.3 per million for women.

5   Informants suggested that if, prior to an X-ray, one smoked about 30 cigarettes or drank milk or condensed milk, it increased their chances of being diagnosed with *mynstof.* The smoke was said to make your lungs *vaal* (pale, white or bleak) while the milk provided a whiteness and looked like calcification.

6   Since the 1980s, official policy has prohibited trade, application and supply of blue asbestos (crocidolite) and brown asbestos (amosite). A host of legislation controls work environments and asbestos-related risks including the Health and Safety at Work Act 1974, the Asbestos (Licensing) Regulations 1983, the Control of Asbestos at Work Regulations 1987, which was amended in 1992, the Control of Asbestos in the Air Regulations, 1990, and the Asbestos (Prohibitions) Regulations, 1992. Amosite and crocidolite were banned in 1985/1986 and chrysotile asbestos in 1999 (Kazan-Allen, 1999).

7   This prolonged process requires that a claimant's employers were insured, although compulsory insurance in the form of employers' liability was only introduced in 1972. Many laggers and other workers were, in fact, exposed to asbestos long before this date. The Financial Services Compensation Scheme is intended to cover those cases where both companies and their insurers have gone out of business and which therefore cannot be sued. The scheme pays 100 per cent if the claim was covered by compulsory insurance, and about 90 per cent for cases of exposure before 1972 that were not covered by compulsory insurance. People can also claim from the

government Department of Works and Pensions which administers two schemes, namely the Industrial Injuries Disablement Benefit and the Pneumoconiosis etc. (Workers' Compensation) Act of 1979. The Industrial Injuries Disablement Benefit is for patients who contracted the disease while in employment after 4 July 1958. The Medical Board assesses each case and only provides compensation when disability is judged to be greater than 14 per cent. This generally excludes cases of pleural plaques and pleural thickening. The Pneumoconiosis etc. (Workers' Compensation) Act provides a fixed rate of compensation (determined from scales for age and degree of disability) for workers whose employers are no longer in existence. Compensation can be obtained for diffuse pleural thickening, asbestosis, lung cancer and mesothelioma. This compensation can also be awarded to widows and other dependants who may claim after the death of the worker (Field Fisher Waterhouse, n.d. Field Fisher Waterhouse, 2006).

8   In order to claim compensation, one had to be able to show evidence of mine employment. As the mines had not kept records and many workers were illiterate, other former employees completed the application forms and vouched for applicants' former employment on the mines.

9   However, research in the Northern Cape noted significant undiagnosed and uncompensated instances of ARDs (Hopley and Richards, 1999).

10 Pleural effusions are a build-up of excess fluid in the pleural cavity, a fluid-filled space around the lungs, that prevent the expansion of the lungs, impair breathing and are extremely painful.

11 Residents of Barking and Dagenham also experienced financial constraints and inadequate medical facilities. Ian Bayley became really sick in 2004 and was permanently confined to bed. Initially, the respiratory nurse came to visit every six months or so, but then financial cutbacks meant that she stopped visiting.

12 Balshem similarly describes this alien-like quality in her work amongst residents of a Philadelphia, US, cancer 'hotspot'. Her informants described cancer as a 'great big thing that's eating up your whole insides' and as being like a 'pacman gobbling up your insides' (1991, p158).

13 See for example Franklin and Ragoné, 1998; Lublin, 1998; Clarke and Olesen, 1999; Rapp, 1999; Becker, 2000; Edwards, 2000; Rosser, 2000; Franklin and McKinnon, 2001; Franklin and Lock, 2003; and Handwerker, 2003.

14 At the end of the 20th century, the Griqua comprised a diverse category of people, historically seen as the 'mixed-race' descendants of indigenous Khoi (nomadic pastoralists), autochthonous San (nomadic hunter-gatherers), escaped slaves, *Boer* frontiersmen, Africans (predominantly Tswana) and European settlers. The name Griqua – which refers to the indigenous Khoi resident in the Cape Colony in the 17th century – was adopted in 1813 when the London Missionary Society persuaded the heterogeneous collectivity at Klaarwater (Griquatown) – then known as and calling themselves 'Bastaards' (literally, bastards) – to change their pejorative name (Halford, 1949; Nurse, 1975; Ross, 1976).

15 According to the President's Commission investigating the needs and demands of the Griqua, 'a Griqua is a coloured and a coloured is a Griqua on the grounds of

the definition that "a coloured is neither white nor Black nor Asiatic", and thus all the other coloureds must be classified in this class' (1983, p76).

16   During the apartheid era, coloured and white workers were classed together with different compensation requirements for African workers (Myers, 1981).

17   Medical examinations are only a contractual requirement when working for big companies or when applying for a licence to strip asbestos. As most of the laggers now work as independent contractors, they are not required by law to have regular check-ups.

18   Perhaps it was for this reason, because acknowledging the problem involved recognizing the part that men have unwittingly played in their families' lack of well-being – albeit precisely because of their desire to provide for the family – that laggers seldom discussed these matters with their wives. Very few men interviewed admitted discussing asbestos dangers with their wives. Conversations around these matters were kept to a minimum and comprised absolutely essential issues such as the outcome of doctor's appointments and legal processes. Very seldom did husbands and wives discuss the implications of a certain diagnosis, the possibility of contracting more severe forms of ARDs or the risks to other family members.

Chapter 6

Reframing Risk: Comparative Framings of Asbestos and Disease

The asbestos debate is over, except for dealing with the legacy issues.
(Trade unionist, South Africa, March 2008)

Asbestos remains asbestos, disease remains disease, regardless of where in the world you are or how your government defines it. (Griquatown resident, June 2009)

Introduction

The previous chapter explored victims' emic and gendered experience of asbestos-related disease (ARD) in comparative perspective, showing how people's experiences and interpretations did not always mirror medical and legal definitions of ARDs. This chapter uses a national and comparative focus to examine policy processes and to explore how asbestos issues are dealt with at national political levels. These national policy processes involve diverse social, technological and environmental considerations and trade-offs, and are frequently challenged by various groupings of activists and social movements seeking to influence national political processes around asbestos issues. Focusing on the UK, South African and Indian contexts, this chapter examines what framings are being applied in national political processes around asbestos and related diseases (compare with Leach et al, 2007, p7). This approach encompasses both societal and scientific understandings of ARDs and risk, and the subjective ways in which these are reinterpreted through its emphasis on framings. Framings pervade governance processes and social protest, producing particular narratives, conceptualizations of risk and uncertainty and, ultimately, particular forms of sustainability and social justice. The comparison between countries and their framings serves to expose the connections between science and politics. As Jasanoff points out, cognitive frames 'impose discipline on unruly events by creating understandable causal relationships, identifying agents of harmful behaviour, and finding solutions that convey a sense of security and moral order' (2005, p24). As is evident in the previous chapters, different governments define the toxic content of asbestos – and the associated risk – differently, interpret ARDs through different medical lenses and use different forms of scientific expertise to position their countries in relation to global asbestos debates. In response, social movements and anti-asbestos activists attempt to assert alternative understandings of asbestos and risk. As demonstrated in previous chapters, these movements reveal interrelated 'global science' and local civil society and community dynamics (compare with McCormick, 2009, p5), yet, as governments and activists frame asbestos concerns in divergent ways, these dynamics play out differently in different country contexts. The chapter is thus an exploration of the ways in which political, scientific, technological and governance processes interact around knowledge of asbestos. In order to explore framings and conceptualizations of risks and uncertainty, the discussion that follows is guided by the following questions: What are the dominant understandings of asbestos among powerful actors and how do alternative visions contest this understanding? What assumptions are included in these dominant framings and what kinds of knowledge, understandings and experience are excluded? And, finally, what implications do these dominant framings have for broader understandings of incertitude such as risk, uncertainty, ambiguity and ignorance? (Leach et al, 2007, p8)

Broadly speaking, two framings of asbestos are, or have been, dominant in all three countries. The first is the framing of asbestos as cheap and modern and as having considerable value, saving lives through its fireproofing qualities. This framing – advanced by the economically powerful asbestos industry – has particular historical resonance, having stemmed from the 1930s when a cartel of multinational corporations emphasized asbestos's positive attributes, downplayed the health risks, invented new uses for asbestos and financed scientific research (McCulloch and Tweedale, 2008). This understanding has retained prominence in many aspects of asbestos governance in developed and developing countries. As demonstrated in previous chapters and discussed further below, South Africa, the UK and India have all experienced periods in which asbestos has been seen as cheap and as a product of modernization. A second dominant framing – perpetuated by both the asbestos industry and governments in all three countries to varying degrees – is the idea that asbestos exposure and the associated risk of disease is controllable. The discussion that follows shows how this particular understanding of asbestos has increased in prominence in countries that have banned asbestos, particularly in relation to in situ asbestos embedded in buildings. These dominant framings are bolstered by legal and medical understandings of ARDs. Here, as the reader is no doubt already aware, there is considerable variation between countries. This has ramifications, not only for the identification, compensation and treatment of disease, but also for how national governments are 'positioned' in relation to social movements, activism and in terms of how people interpret dominant framings. As the following section will show, although dominant framings in all three countries rely on scientific and statistical probabilities of risk, considerable variation occurs through the different positioning of countries in relation to global asbestos debates, countries' varied interpretations and practices of democracy and the exposure of different populations to asbestos.

Pleural plaques and the control of risk: Dominant framings in the UK

The two interrelated framings which dominate UK debates are first, that all asbestos is dangerous but controllable, and second, that pleural plaques do not indicate that these sufferers are at risk of further ARDs and are, instead, undermining UK societal and civic values in their campaign for compensation (De Saulles, 2006a). These ideas are perpetuated by high-ranking judicial actors in the UK House of Lords, powerful industrial and insurance lobbies, scientific specialists who participate in government inquiries and by government bureaucrats through diverse occupational health and safety policies (discussed in detail below). The notion that asbestos is dangerous but controllable has a long history in the UK. Although the possibility of contracting ARDs was well-established by the 1930s, companies maintained that controlled production methods allowed them to protect workers. Thus, throughout the era in which asbestos companies operated in the UK (1898–1999), emphasis was placed on controlling fibre counts, on scientific procedures of measuring fibre size and sample, and correlating exposure to risk. Tweedale documents, for example, the considerable energies invested in determining an 'acceptable' threshold level of exposure alongside widespread failures to maintain health records, implement safety procedures or conform to safety regulations. When, in the 1980s and 1990s, social mobilization challenged this dominant framing by raising awareness of the prevalence of ARDs and questioning biomedical notions of ARDs (McCulloch and Tweedale, 2008; Tweedale, 2008), the UK government responded by banning crocidolite and amosite, arguing initially that blue and brown asbestos were the most carcinogenic and that chrysotile asbestos exposure could be controlled. It took a further ten years of mobilization before white asbestos was banned. It is clear that, during the 1980s and 1990s, the dominant narrative of asbestos as quintessentially modern, safe and cheap – as put forward by the asbestos and insurance industries – was supported by the notion that risk could be controlled.

Because vast amounts of asbestos had been used[1] and were already embedded in buildings and other urban structures, the 1999 banning of all forms of asbestos only partially challenged this framing. UK health and safety regulations, which shifted from a focus on asbestos manufacture to the management of in situ asbestos, continued to emphasize that exposures are controllable. This official government framing draws on scientific research showing that domestic exposures to asbestos are relatively low (Price and Ware, 2008; Burdett et al, 1989). For example, Burdett et al's examination of buildings with sprayed asbestos, buildings with warm air heaters containing asbestos and buildings without asbestos found that asbestos was not dislodged by air currents (1989; see also Le Guen and Burdett, 1981).[2] In keeping with these conclusions, Peto argued that 'Average fibre counts in contaminated buildings, at least in the UK, are usually less than 0.001fibre/ml, and the corresponding predicted lifelong risk is of the order of 1 in 100,000 for 10 years' occupancy' (1989, p466). This led scientists to suggest that asbestos removal was unnecessary (Peto et al, 1995, p539; Peto, 1989; Abelson, 1990; Chesson et al, 1990; Pearson and Sims, 1992; Craighead, 2008) and the UK Health and Safety Executive (HSE) to advocate '*in situ* management' of asbestos, arguing that it is best left undisturbed if in a stable condition (Gravelsons et al, 2004).[3] This policy measure – and the underlying framing of risk as controllable – is based on scientific calculations of risk.[4] Guidotti (1988), for instance, estimates that an individual occupying a building with in situ asbestos has a 0.003 per cent chance of contracting lung cancer and negligible risk of mesothelioma. Pearson and Sims similarly predict that 'the level of environmental lifetime risk from exposure to airborne asbestos is probably of the order or 1 in 100,000 or even lower' (1992, p379).

Given the framing of asbestos risk as controllable, conditions for working with asbestos are dealt with in the Control of Asbestos Regulations 2006. In summary, these regulations specify that if one suspects the presence of asbestos, the site should be tested. If the tests are positive but 'low risk', specialized equipment and techniques can be used to remove the asbestos or, if 'high risk', specialized teams should be contracted for its removal. The regulations specify legal fibre concentrations, exposure levels and exposure time periods, as well as detailing appropriate equipment and techniques for controlling fibre concentrations.[5] These regulations focus on non-domestic exposure to asbestos, requiring all employers and the self-employed to minimize or prevent asbestos exposure and identify who has a specific duty to manage asbestos (Burdett, 2006). In terms of domestic premises, landlords' responsibilities and obligations are detailed in the Defective Premises Act 1972, which specifies that tenants should be protected from identified hazards and problems. In addition, since May 2006, the HSE has mandated that all UK business property managers record the presence of asbestos in an asbestos register for non-domestic premises. Landlords' legal duties are not to remove all asbestos, but to manage it in public areas. This requires an assessment of the condition of the asbestos and of the likelihood of fibres being released, an asbestos register and a management plan (that specifies whether asbestos is to be sealed, encapsulated or removed) followed by regular inspections (HSE, 2006). The primary recommendation of the HSE is that, unless corroded and of poor quality, all asbestos should be encapsulated and managed.

This framing – that proposes scientific calculations of risk can be trusted and that these calculations point to negligible risk of exposure – is also perpetuated in the current political and legal wrangling regarding the status of pleural plaques. The dominant framing here is that asbestos victims' demand for compensation poses a threat to societal values. As discussed in Chapters 2 and 5, the medical and legal status of pleural plaques has, since 2007, changed from a recognizable and compensatable ARD to an 'irrational' form of anxiety. A sustained social movement has protested since the House of Lords' decision in October 2007. In response to political pressure and social protest, Prime Minister Gordon Brown initiated a consultation process on the pleural plaques issue. In addition, the government requested that the Industrial Injuries Advisory Council (IIAC) – an independent statutory body that advises the UK government, and particularly the Secretary of State for Social Security – review pleural plaques in relation to the Industrial Industries Disablement Benefit scheme, in order to ensure that government decisions are based on the 'best available current medical evidence' (Ministry of Justice, 2008, p15). In July 2009, the IIAC released its position paper on pleural plaques.

The IIAC began with the legal requirements for the prescription of occupational disease, which specify that occupational diseases must pose a 'recognized risk' to workers and that the relationship between disease and occupation 'can be established or reasonably presumed' (IIAC, 2009, p10). This means that, in cases where a disease might not be confined to occupational circumstances, epidemiological evidence has to demonstrate that occupational exposure doubles the risk of contraction. This 'doubling of exposure', the IIAC argues, 'is not arbitrary', but stems from the definition of hazardous as a 'doubling of risk'. The IIAC report recognizes that pleural plaques are under-identified and that conservative estimates place pleural plaques as the second most common form of occupational lung disorder. It also acknowledges that the increased incidence of pleural plaques is related to heavy exposure to asbestos. Ultimately, however, the report argues that pleural plaques neither change the lung structure nor curb lung expansion. Indeed, where pleural plaques do seem to have caused difficulty in breathing or other symptoms, the IIAC suggests a range of possible alternative explanations: other, undetected and underlying cases of lung fibrosis, pleural thickening, age, smoking, personal anxiety and extensive calcification (2009, pp25–28).[6]

The IIAC thus suggests that pleural plaques are not a '*cause* of cancer', although the presence of pleural plaques has been statistically 'linked with a greater *future risk* of these cancers' (IIAC, 2009, p28, original emphasis). 'This relation is one of a *marker*, not a cause. Plaques are a marker of exposure to the causal agent, asbestos' (IIAC, 2009, p29, original emphasis). This means that, as Dr Dunne and others argued in Chapter 2, 'any increase in risk in those with pleural plaques arises because they have been exposed to asbestos, not because they have pleural plaques' (IIAC, 2009, p35). In conclusion then, the IIAC argued that pleural plaques created anxiety because workers and patients 'confused' pleural plaques with other ARDs and because they were concerned about the future risk of cancer. Thus, pleural plaques do not cause lung impairment, do not represent a hazard (in the sense of doubling the risk for occupational exposure) and should not be considered for compensation as the Industrial Industries Disablement Board's intention is to 'compensate *actual* disablement, rather than *future risk* of disablement or *health anxiety*' (IIAC, 2009, p38, original emphasis). This report by the IIAC is significant because it establishes the groundwork for the government's long-delayed decision on the House of Lords ruling. Although Gordon Brown reassured campaigners of his intention to seek justice, his consultation concluded that compensation for pleural plaques is, in effect, 'giving compensation payouts before a disease develops'. This decision is based on the Chief Medical Officer and the IIAC's view, emphasized 'more firmly than ever before' that pleural plaques are 'not harmful' (B. Roberts, 2009). This disjuncture between government empathy and scientific proof of risk, coupled with endless consultation and delays, can be seen as a classic case of 'organised hypocrisy' (Brunsson, 2003). It is a means of prevarication that – in the face of an election – allowed the Labour government to appear to be taking seriously the concerns of workers, others exposed to asbestos and its traditional support base without alienating commercial interests. For the activists and victims concerned, however, this hypocrisy positioned the Labour government 'against' the people as the following statements from asbestos victims indicate. 'You're not in it for us, you are in it for yourselves! That is what working people think of the Labour Party,' said one asbestos victim. 'Our party is no longer a party of ordinary people', said another, 'we write to Jack Straw and Gordon Brown and we get trite back.'

This dominant framing of pleural plaques as benign, symptomless and unrelated to the risk of contracting other ARDs is underlain by a notion that asbestos claimants are threatening UK society and its principles. The language used in the court cases described in Chapter 2 incorporates ideas of unwarranted compensation, irrational claims and personal character defects. Thus, the court cases detail fears of increased liability, scam vans and dubious claims. These dominant notions – of unjustified compensation – are reinforced by the IIAC report, which states that 'any losses (of lung function) fall well short of the compensatable level of disability within the Industrial Injuries Scheme defined previously in relation to chronic obstructive pulmonary disease' (IIAC, 2009, p5). The notion that the compensation of pleural plaques is illogical and violates fundamental social values is, as De Saulles (2006c) points out, a product of 'tort tales' that – like framings – use rhetorical techniques[7] to encapsulate simplified versions of reality.

In contrast to the dominant view that pleural plaque sufferers are 'on the make', a complex set of interrelated historical and present-day factors have shaped the UK situation. These include: employers' failure to admit the gravity of asbestos exposure and their reluctance to take health and safety seriously; workers' patriarchal values, their tough, self-reliant image and their determination not to reveal weakness; the provision of inadequate data to previous governments; poor surveillance and insufficient factory inspections; the high profile of other occupational diseases; the close relationship industry has with some occupational health specialists; the agreement between government, medical expertise and industry that asbestos production was necessary for economic growth and for safety (provided through its fireproofing qualities); and ironically, the presence of legislation which enabled people to feel that adequate provisions were in place (Tweedale, 2000). There is, within the House of Lords' legal framing of the pleural plaque cases, no recognition of these interwoven and complex factors. Instead, as De Saulles has shown, a focus on individual cases of exposure has meant that the Law Lords failed to see the 'big picture'. As a result, pleural plaques have become part of a broader legal debate 'centred around the question of whether or not too many people are actually making a claim' (De Saulles, 2006c, p306). As argued in Chapter 2, there is clear evidence that pleural plaque sufferers are not rushing to scam vans and heedlessly seeking compensations, but are reacting in accordance with their very particular socio-economic positions and intimate knowledge of ARDs. Ironically, however, the medical evidence that there is an 'epidemic' of asbestos claims has worked against the claimants. This framing, that there is a danger of extensive – and unwarranted – litigation, shifts the focus from historical and socio-economic considerations to individual accounts of pleural plaques. This, in effect, allows the courts to overlook attempts by insurance companies to avoid compulsory insurance, to limit their liability and to avoid paying compensation through delaying tactics (Tweedale, 2000). Corporate interests – well positioned to benefit from negotiations with governments because of their experience, technical knowledge and financial resources (Murphree et al, 1996) – also emphasize their close affiliations with the judicial system, using rhetorical devices to portray themselves as like-minded individuals concerned about economic growth, business and societal well-being and distancing themselves from the claimants. This facilitates the framing of pleural plaques, not as an occupational disease for which victims might claim compensation, but as a greedy desire for money that will ultimately undermine the judicial system (De Saulles, 2006c).

There are thus two dominant framings within the UK, perpetuated by government agencies, powerful judicial actors, some medical specialists and some researchers, which centre on the scientific notion of risk. The first is one of calculated scientific risk in which it is argued that asbestos is dangerous but exposure is controlled through legislation and encapsulation. The second dominant framing suggests that people who suffer from pleural plaques experience a risk that they will suffer from other ARDs, but this risk is not increased by the presence of pleural plaques. Because sufferers misinterpret this risk, their pleural plaque claims will lead to a compensation culture and to the subsequent degeneration of UK societal values. The use of scientific approaches to determine the extent of risk also forms the basis for dominant framings in the Indian context, albeit, as the following section shows, with significant deviations from the UK case.

Risk-free asbestos: The dominant Indian position

Science and risk, so evident in the dominant UK framings of asbestos, are also the underlying basis for India's dominant framing of asbestos. Here, the primary approach perpetuated by the government and the asbestos industry has been that asbestos risks are non-existent, or, if there are minimal risks, these are controllable through production methods. As demonstrated in Chapter 4, this dominant framing has been facilitated by the 'lack' of Indian evidence of ARDs and the legalization of chrysotile, resulting in a highly scientific debate regarding the dangers of asbestos. The Indian government – under pressure to ban asbestos – has conducted its own research on chrysotile asbestos (DGFASLI, 2006; Dutta, 2008a, b). In 2004, the Indian Ministry of Chemicals and Fertilizers commissioned the National Institute of Occupational Health (NIOH) to research the health and environmental hazards of chrysotile asbestos in preparation for the international Rotterdam Convention. This study, based on current workers at asbestos cement factories which have been operating for 10–15 years, uses sophisticated techniques to count only scientifically defined 'fibres', which have an aerodynamic diameter greater than 5 micrometres in length and a width smaller than 3 micrometres (Lemen, 2008). Furthermore, this scientific approach overlooks latency periods and asbestos particles not technically defined as fibres.[8] Although this research has produced some evidence of 'impaired lung function', this has not been sufficient to challenge the dominant framing of asbestos as safe. Instead, coupled with the official

Indian definition of ARDs, this demonstrates how the use of scientific formulae and calculations of risk can minimize the possibility of statistically linking asbestos exposure to disease. This government framing of asbestos as risk-free has significant economic implications. In India in 2006, Indian-owned Everest representatives firmly clung to the dominant framing of asbestos as safe and of the risks as controllable. They argued that the company could make non-asbestos products, but that they 'can't see the difference [or point]' in doing this because health risks are controlled during production, substitutes cost more and there is no market for non-asbestos products. Everest thus sees anti-asbestos mobilization as stemming from a deficit of scientific knowledge, an ignorance of health and safety measures and from lawyers' avarice. As such, Everest sees itself as accountable to the state and its investors through a version of corporate responsibility that promotes cheap, accessible, asbestos-ridden products in order to help India 'progress'.

In India, public participation is used to reinforce this dominant framing of asbestos. Before manufacturing units or factories can be constructed, Environmental Public Hearings (EPHs) are held. These are open to anyone and local people likely to be affected by the construction are encouraged to participate. These situations, however, tend to be advantageous to industrial actors who have greater experience, technical knowledge and financial resources (Murphree et al, 1996, p448). Strong relationships – or what McCormick (2009, p3) terms an 'iron triangle' – between companies, government and research use the 'scientization of politics' to exclude citizens from participatory decision-making. This process, and the use of scientific expertise, is particularly evident in the case of New Sahyadri Industries Ltd (NSIL), which wished to establish an asbestos unit in Surat, Gujarat. NSIL offered guarantees of strict precautionary measures, state of the art technology and evidence from other NSIL plants. During the EPH, it assured local participants that there is 'absolutely no problem regarding pollution or occupational health' because the company uses only white (chrysotile) asbestos: 'as per various scientific studies carried out the world over, there is clear evidence that when only Chrysotile (White) Asbestos is used in a controlled manner, there is no risk to humans or to the environment' (letter to Environmental Public Hearing Committee chairman, 24 January 2006). In closing the EPH, a member of the Environmental Public Hearing Committee (EPHC) commended the local villagers on their environmental awareness, berated the local women for not being present, and thanked all the participants for their 'active participation'. NSIL was subsequently provided with a 'No Objections Certificate' from the Ministry of Environment and Forests, allowing it to go ahead with the construction of the asbestos cement plant,[9] despite local villager protests, written submissions and signed petitions.

Citizen participation, in the context of this framing of asbestos as safe and of ARDs as negligible, is thus highly symbolic and does not provide a policy forum in which community representatives can voice alternative framings or challenge dominant viewpoints. It is participation without 'cognitive representation' (Visvanathan, 2009, p156). Instead of being able to engage in scientific debate with local residents and acknowledging a 'plurality of knowledge systems' (Visvanathan, 2009, p156) that differently evaluate risk, factories received environmental clearance based on their assurance of scientific techniques and their intended close monitoring of environmental conditions and residents' health. Such participation reinforces India's institutional framing of asbestos which is modelled on the idea of an ignorant public trusting governments and scientific experts to manage risk and ensure citizens' safety (Zavestoski et al, 2004). In contrast to this Indian example, and as examined in the following section, South African dominant framings use science to bolster the relationship between ordinary people and government understandings of risk.

Using science for the people: South African dominant framings

The South African government has been able to use the asbestos issue, and its framings thereof, to position itself as a 'caring government' working in partnership with civil society. This stems from a reliance on scientific calculations of risk used – in contrast to UK or Indian government framings – in conjunction with government awareness of historical precedent, of socio-economic circumstance and of political change. In 1998, following the transition from a racist, apartheid state to the democratic, rights-based government of the African National Congress, a National Asbestos Summit was convened by the Portfolio Committee on Environmental Affairs and Tourism, which put asbestos on the national agenda. Trade unions described the government's attention to asbestos issues as an 'example of a caring government' partnering with civil society (NUM, 2007). Participants at the National Asbestos Summit (including scientists, medical doctors, government officials, trade unionists, industry representatives, local activists, health and safety executives, South African ministers and international experts) readily accepted the scientific argument that all asbestos was carcinogenic and called for a ban on all forms of asbestos manufacture, including chrysotile. This resolution was supported by members of the South African Parliament who saw it as their role to 'make sure that the conclusions of the [National Asbestos] Summit do not remain on paper but are translated into action' (Jerry Ndou, South African MP, 1998). As a result, and despite some initial reluctance (McCulloch and Tweedale, 2008), this resolution was endorsed by the South African cabinet in 1999 and, after a period of planning and deliberation,[10] March 2008 saw the introduction of the 2008 Asbestos Regulations prohibiting the use, manufacturing, import and export of all forms of asbestos and asbestos-containing materials.[11] For many South Africans, this ban was seen as a 'moral and political obligation' of the government to victims and labourers and as a formal recognition of the knowledge that all ARDs are preventable (Gibson, 2007a, p14).

Prior to the end of apartheid and the National Asbestos Summit, South African industrial scientists had argued that asbestos exposure was no more dangerous than coal smoke and car accidents. This was part of a long history of collusion between industry and the apartheid government, which included infrequent inspections of remote mines, the granting of special privileges to asbestos mines, inadequate health and safety requirements and the labelling of alternative viewpoints about asbestos as hazardous as 'communist threats' (McCulloch, 2002). However, community organizations such as Concerned People Against Asbestos (CPAA, described in Chapter 3) and trade unions (supported by 'progressive' doctors) readily adopted international trends advancing the idea that all asbestos was dangerous. In contrast to Indian trade unions, South African trade unions were determined not to facilitate workers' exposure to toxic products. At the 1998 National Asbestos Summit, they demanded that employers scale down their production of asbestos-containing products, allowing retirement and natural attrition to reduce the numbers of workers exposed to asbestos. They also insisted that suitable substitutes be found, so that companies could phase out their asbestos production rather than simply closing down. As one trade unionist recalled: 'It was very serious, we knew that employers would retrench anyway when under pressure, [so we debated] do we wait and let people be exposed to asbestos or do we tackle it head on? We demanded that companies move out of asbestos production without job losses.' Although several companies used these demands to shut down production plants, Congress of South African Trade Unions (COSATU) representatives saw the struggle over asbestos and the implementation of the ban as 'a victory for trade unions and for society as a whole. Trade unions have been losing members [as people die of asbestos diseases] who trust trade unions on occupational health matters.'

COSATU's struggle to ensure that South Africa formally recognized the hazardous nature of asbestos, yet simultaneously sustained workers' livelihoods through careful phasing out of asbestos, also shaped some industrialists' viewpoints. The formerly Scandinavian-owned manufacturing company Everite was well-positioned to receive and respond to this framing. Having operated in South Africa since the 1930s (McCulloch and Tweedale, 2008) with a sophisticated public relations and media component, Everite had spent many years denying the possibility that embedded asbestos might be hazardous to human health. Chapter 3 documents the manner in which, during the 1980s and 1990s, Everite used scientific evidence, sophisticated ideas of fibre counts and international standards to refute the CPAA's concerns regarding asbestos cement roofs. By mid-2000, Everite faced a number of legal cases from South African asbestos victims living near Everite factories (Cullinan, 2002), as well as public exposure of the company's callous indifference and suppression of knowledge regarding ARDs (Braun and Kisting, 2006). It became highly attuned to the changes in South African opinion, as well as being conscious of how Scandinavian asbestos bans and international court cases had affected the parent company in the 1970s (Kazan-Allen, 2002). These insights led Everite to reassess its position: after years of insisting that asbestos was safe, cheap and modern, Everite stopped seeing anti-asbestos campaigners as adversaries. This 'helped the industry come to terms with the inescapable fact that the future of ... its raw materials was uncertain' (van Zyl and Gibson, 1987, p7). Linked to a broader national reassessment of asbestos and a dominant framing which stressed its hazardous nature, Everite created a forum for South African businesses to consider alternatives, launched internal health and safety campaigns and initiated an employee compensation scheme. Everite public relations personnel engaged in national and international policy processes, working alongside trade unions and anti-asbestos activists and – paradoxically – supporting an asbestos ban.[12] This South African company, although ultimately driven by economic incentives, was – through its reframing of all asbestos as hazardous and of production methods as creating uncontrollable risk – able to ride a groundswell of public opinion, retain some accountability to its workers and to position itself strategically within asbestos-related political processes in South Africa.

Asbestos is embedded in countless South African houses and other structures. In keeping with dominant framings in India and the UK, South Africa is confident that asbestos already embedded in building structures does not pose a risk despite, as described in Chapter 3, the occasional removal of these products or heightened media concern about in situ asbestos (see, for example, Joubert, 2007). Drawing on scientific evidence emphasizing the low risk of in situ asbestos, the 2008 Asbestos Regulations do not specify the removal of asbestos embedded in houses and other infrastructure, but recommends that these ultimately be replaced (Department of Environmental Affairs and Tourism, 2008a). Echoing the UK's position of minimal risk, the Department of Environmental Affairs and Tourism, under whose auspices the regulations were promulgated, comments: 'The Department is satisfied that there is no undue risk to the occupants of houses that are fitted with these materials' (Department of Environmental Affairs and Tourism, 2008b, p1). South African trade unionists have accepted this position as they seek to balance their asbestos concerns with considerations of poverty and unemployment. They thus find themselves asking, 'how can we demolish houses while others don't have homes to live in?' In addition, replacing asbestos roofs can expand the problem: not only because, as shown in Chapter 3, residents do not take adequate precautions, but also because asbestos roofs or other cement products are seen as highly 'desirable' building materials. Unless steps are taken to ensure that the asbestos-containing cement is immediately removed and officially disposed of, it often 'disappears' to be used in someone else's home (Phillips et al, 2007). Nonetheless, this statistical conceptualization of risk is currently undergoing revision within the dominant framing of asbestos. Much of this reframing stems from increasing awareness that South Africa's incidence of ARD contagion is not decreasing. In addition, given the mine closures and long latency periods associated with asbestos, the fact that people in their early 20s are contracting mesothelioma and other ARDs has raised new questions about the hazards of exposure (Visagie, 2008). Explanations have now focused on environmental conditions rather than on objective, but abstract, scientific calculations of risk with increasing recognition that people interact with their physical environments and can disturb embedded asbestos fibres.

South Africa's introduction of the new legislation and asbestos bans has coincided with revived environmental concerns (initiated because rural residents were still being diagnosed with ARDs despite the cessation of mining for more than 20 years) about secondary asbestos pollution. Dealing with the 'after-effects of primary and secondary pollution' from asbestos has become both a government priority and a challenge. In addition to the continued rehabilitation of former mines, the towns of Griquatown, Prieska and Kuruman have benefited from several government training schemes and sustainable development projects between 2006 and 2008. Run by the Department of Minerals and Energy Affairs, these have sought to impart information and skills in processing semi-precious stones, mining and the hazards of asbestos and on the rehabilitation of asbestos mines. These efforts of coordination have also extended beyond local communities with workshops organized in towns such as Prieska to share information across diverse stakeholders, including the Medical Bureau for Occupational Diseases (MBOD), provincial government departments, non-governmental organizations (NGOs), asbestos relief trusts such as the Asbestos Relief Trust (ART), community-based organizations and social activists (Saaiman, 2008).

The Department of Minerals and Energy Affairs, tasked with reducing the environmental liabilities and pollution, decided in 2007 to focus its activities on asbestos mines because of the associated health risks (Department of Minerals and Energy, n.d.). As discussed in Chapter 3, much of this rehabilitation has been done in association with local stakeholders. In recent years, it has also been complemented by awareness programmes, job creation, project implementation committees and public meetings in order to ensure a sense of community ownership (Government Communication and Information System, 2009). This concern with environmental pollution and secondary exposure to asbestos has also resulted in a number of studies. In 2008, the Department of Environmental Affairs and Tourism initiated research in heavily affected areas (Visagie, 2008). This intensive research, which examined more than 2000 samples from 40 communities,[13] showed that local communities experienced extensive environmental asbestos pollution and that, despite the rehabilitation of former mines, the communities' built environments had not 'been adequately assessed to determine the limits of environmental asbestos contamination' (Jones cited by Smit, 2009). A follow-up study was undertaken in Prieska which examined many housing samples and confirmed that, despite the initiatives to replace asbestos roofs described in Chapter 3, many houses continue to be contaminated by asbestos. Schools form a particular area of concern with asbestos pollution both within school buildings and playgrounds and on the routes to and from schools (Smit, 2009).

These research results were presented to the Northern Cape Executive Council – which currently oversees an intergovernmental committee that brings together the departments of health, environmental affairs, social services and local government, in order to raise the profile of asbestos issues in relevant national departments. In other areas where South Africans have been heavily affected by asbestos mining and secondary pollution,[14] the provincial government departments have requested that the Department of Environmental Affairs and Tourism coordinate their response. This involves assessing the financial implications for a cogent rehabilitation programme and approaching the National Treasury for this funding (Saaiman, 2008). As South Africa is finding out, the problems associated with managing asbestos, removing it and simultaneously protecting residents are not easily dealt with, be it in policy or practice. The dominant government framing of asbestos is one that draws on scientific and technological understandings of risk, but does so in relation to social and environmental considerations rather than in isolation. In the sense that it recognizes the ways in which people live and interact with their built environments and has empathy with their concerns about health and well-being, it represents a departure from the UK government and official Indian policy understandings of asbestos.

The domination of science and risk

In all three countries, official framings of risk have been based on expert research, which relies on formulaic, scientific calculations of likelihoods and outcomes (compare with Jasanoff, 2009; Stirling, 2009). As is evident in the above section and in Table 6.1 below, dominant framings tend, with the exclusion of the South African case, to focus on biomedical and technical concepts while dismissing broader social and environmental considerations. This medical scientific information is mediated by experts who act as 'brokers of information' serving government institutions, corporations and – on occasion – activists and social movements (McCormick, 2009, p29). Their research feeds into official policy, suggests technical or legal solutions, and facilitates the 'scientization' of economic and political issues. Nonetheless, this country comparison reveals some of the different ways in which expert framings influence science, politics and technology. In the UK, the dominant ideas are that all asbestos is lethal and banned, that in situ asbestos is controlled through health and safety. In relation to pleural plaques, the scientific concept of risk demonstrates that sufferers are not more likely to contract other ARDs. Shoring up the legal defences against people wishing to claim compensation for pleural plaques is thus also a means to protect society from 'unwarranted' litigation by 'irresponsible' and 'socially unworthy' claimants. Ironically, however, this approach and the delaying tactics displayed by the UK government in the face of repeated calls to overturn the House of Lords' ruling, resulted in positioning the Labour government 'against' the people. In India, the government approach to asbestos is one that was previously popular in both South Africa and the UK, namely that asbestos is cheap, modern and that the minimal risk is made safe through the implementation of production techniques. Science is the infrastructure for this dominant framing that sees the notion of risk as stemming from western imperialism. Similar echoes are apparent in the South African apartheid government's understandings of asbestos, when asbestos mines held a 'special place' in the nation and when risk and disease were seen to be inspired by 'communist' ideals. Now, however, the South African government approach to asbestos, supported by powerful stakeholders in industry, trade unions and by community organizations, is similar to that of the UK, with the one substantive departure being South Africa's recognition of and compensation for pleural plaques. As discussed in Chapters 3 and 5, pleural plaques are seen as precursors to other ARDs and South Africans can claim compensation through the MBOD, the Disability Grant or, more recently, through the ART and other relief trusts.[15] This recognition of people's suffering, the labelling of asbestos as a problem, the official government support given to the Cape plc case and the governmental concern regarding environmental rehabilitation and secondary asbestos pollution, alongside the recent banning of asbestos, have allowed the South African government to position itself as a government 'of the people'.

As the above section shows, although all three governments relied primarily on scientific calculations of risk, their framings were also influenced by other political processes and powerful actors. Thus, although official interpretations emphasized the scientific basis from which policy stemmed, science alone did not

Table 6.1 *Dominant framings and primary actors in the UK, South Africa and India*

	Dominant Framings	**Primary Actors**
UK	All asbestos is dangerous but controllable. Asbestos embedded in buildings is low risk.	UK Health and Safety Executive, scientists researching asbestos exposure in buildings.
	Pleural plaques benign, symptomless and unrelated to the risk of contracting other ARDs. Their claims undermine UK societal and civic values.	Prime Minister's inquiry, IIAC, Law Lords in the House of Lords, corporate interests and insurance companies.
South Africa	All asbestos is carcinogenic and must be banned.	National Parliamentary Portfolio Committee on Environmental Affairs and Tourism, scientists, medical doctors, government officials, trade unionists, industry representatives, local activists, health and safety executives, South African ministers and international experts.
	Asbestos substitutes need to be found without job losses.	Trade unions, select asbestos industries such as Everite.
	Asbestos embedded in buildings is low risk.	Department of Environmental Affairs and Tourism, trade unionists.
	Asbestos mines need to be rehabilitated because of environmental concerns.	Local stakeholders and communities, Department of Minerals and Energy Affairs, provincial government departments, NGOs, asbestos relief trusts, social activists, Northern Cape Executive Council.
India	Chrysotile asbestos is safe. Therefore asbestos risks are minimal or non-existent, and can be controlled through production methods.	Indian government, in particular the NIOH, asbestos industries including Everest and NSIL.

determine how asbestos issues were interpreted and responded to. Questions of material resources, political interest and, in particular, political context strongly shaped how science and risk predictions were interpreted. Powerful insurance lobbies and industrial representatives influenced governments to interpret science in particular ways. With the exception of South African dominant narratives, which sought to include social and environmental narratives as put forward by local actors, these 'scientific' decisions elide normative and political considerations. As is evident in both the UK and Indian examples, this type of state administration of expert knowledge has led to undemocratic governmental practices and the marginalization of citizens (compare with Jasanoff, 1996, 2009).

Dominant framings are not the only framings in circulation, however. The understanding of workers, ARD victims, activists, trade unionists, some lawyers and academics challenge these framings of ARDs, risk and appropriate policy regulation. As McCormick (2009) has pointed out, activists often seek to contest large-scale financial and government interests, to advance alternative understandings of the problem and to undermine industry. As movements and activists confront dominant framings, they engage with science and critique it on its own terms. But challenging dominant framings also involves exposing how economic and political interests might shape government regulatory approaches, drawing on other, non-scientific understandings of the situation and highlighting environmental impacts. These are attempts to broaden the policy and regulatory frameworks beyond science and technology by inserting social and environmental dimensions of asbestos, focusing on how people feel and respond to the presence of asbestos or the diagnosis of disease and how they would like to see asbestos addressed in their physical environments. These alternative understandings do not therefore rely solely on scientific knowledge and on calculations of risk. Rather, as shown in Chapter 5, they are often based on people's subjective and practical experiences and on 'embodied' knowledge. In addition, alternative understandings of causality, disease and appropriate policy emphasize other forms of incertitude. In addition to relying on statistical conceptualizations of risk, alternative framings seek to broaden the parameters of the debate, stressing the importance of uncertainty (where outcomes are not statistically calculable), ambiguity (with different categories of people prioritizing different dimensions of the problem or different concerns) and ignorance (where the outcomes are not yet known) (Leach et al, 2007). The following section examines the UK, South African and Indian alternative framings around asbestos issues, asking to what extent these framings are shaped by notions of risk, uncertainty, ambiguity and ignorance, what kinds of issues they emphasize and how this shapes their impact on national political processes.

Alternative framings and social mobilization

Anxiety regarding occupational and residential asbestos hazards: Alternative UK framings

Political activism, combined with high-profile ministerial meetings and media campaigns, has raised the profile of asbestos on the UK national agenda. Trade unions, asbestos sufferers, victim support groups and others have, since the

House of Lords' decision in October 2007, demonstrated their disappointment and disagreement, seeking actively to overturn this ruling. Trade unions, including the earlier described Britain's General Union (GMB), have aligned and 'initiated parliamentary lobbies, held political briefings, coordinated postcard and letter writing campaigns and had the pleural plaques issue raised at the highest levels of government' (Kazan-Allen, 2008, p3). Lawyers and medical specialists, experienced in dealing with asbestos cases and representing pleural plaque sufferers, have produced written submissions demonstrating flaws in the House of Lords' reasoning. Union reports have pointed to failures to regulate and monitor the presence of asbestos in social housing. Members of Parliament (MPs), pressurized by their constituencies, have submitted 'Presentation Bills' seeking to reverse the House of Lords' judgement, and grassroots asbestos campaigners and trade unions have met with the former Prime Minister, Gordon Brown in private consultations. These activities have been accompanied by media campaigns revealing, for example, the poor management of asbestos in the Houses of Parliament and the potential exposure of MPs, a sustained asbestos in schools campaign demanding that asbestos be better dealt with in schools, fringe meetings at trade union congresses, a collection of Early Day Motions dealing with asbestos issues (Kazan-Allen, 2009a), the delivery of signed petitions to Downing Street and the creation of a National Centre for Asbestos Related Diseases. Some MPs have sought official ways of highlighting the need to address asbestos issues and of putting pressure on the government. Andrew Dismore – a trained lawyer specializing in workers' injuries and an MP until 2010 – introduced a Private Members' Bill: the *Damages (Asbestos-Related Conditions) Bill*.[16] If made into law, this Bill would have allowed pleural plaques to be treated as actionable personal injury and people would have been able to claim compensation from insurers regardless of the degree of impairment. In November 2009, the Bill had passed through all the relevant stages in the House of Commons. It had a Second Reading in the House of Lords on 5 February 2010 and reached Committee Stage on 5 March 2010, but as the parliamentary session was prorogued in April 2010, the Bill cannot progress further. Three alternative framings are evident within this large, diverse and ongoing body of protest. These are first, a reframing by laggers and other sufferers of pleural plaques from benign and symptomless to an occupational disease and precursor of other ARDs; second, a heightened awareness from unions and people in the educational sector that in situ asbestos might be lethal and uncontrollable; and third, a reframing of anxiety.

The ways in which laggers – as a particular category of affected people – have sought to challenge the dominant framing of pleural plaques as benign and therefore unworthy of compensation has been dealt with at length in Chapter 2. Between 2007 and 2009, these protests became more energized as laggers aligned themselves with other UK anti-asbestos activists. Their campaign has been complemented by an awareness of how asbestos has historically been concentrated in certain areas and how this has particular effects, not only for the victims, but for the broader communities. Thus, some MPs have argued in parliamentary debates that pleural plaques are a regional issue involving lagging, shipbuilding and railways. In keeping with this alternative understanding of asbestos issues has been the Scottish Parliament's recognition that people with pleural plaques have a far greater risk of developing mesothelioma than members of the general population, and of the legacy of asbestos in its shipbuilding industry. In June 2008, the Scottish Parliament introduced the Damages (Asbestos-Related Conditions) (Scotland) Bill which specified that damages could be claimed for pleural plaques, symptomless pleural thickening and symptomless asbestosis (Kazan-Allen, 2008). The primary intention of the Bill is to ensure that asymptomatic pleural plaques 'is actionable in Scotland' because of the role asbestos played in its industrial past. This is, in effect, a recognition of difference, and a highlighting of the fact that the vast majority of ARD victims come from working-class backgrounds located in former industrial heartlands. As Tweedale wrote in 2000, when the relative benefits of using asbestos were still being disputed, the 'debate has taken on overtones of class warfare. This is hardly surprising as, above all, asbestos-related diseases were linked to social class. The majority of sufferers were working-class people – usually manual labourers – and their "masters" rarely developed asbestos disease' (2000, p286). This alternative narrative demonstrates the prioritization of different concerns in the asbestos debate, or ambiguity. Rather than emphasizing abstract notions of risk and disease so evident in the UK dominant framings, it is a rejection of biomedical statistics. Instead, this framing focuses on who is affected by asbestos exposure and seeks to bring social and environmental considerations to the forefront. This also shifts the debate from whether certain individuals are justified in claiming compensation to a position which recognizes that certain areas and specific categories of people are more likely to bear the brunt of asbestos exposure and risk. This alternative approach also poses questions about individual and collective responsibility (compare with McCormick, 2009, p55).

The second challenge to the dominant UK framing of asbestos focuses on the HSE's confidence that, through legislation and safety procedures, in situ asbestos is adequately dealt with. This challenge comes primarily from the unions, especially the Union of Construction, Allied Trades and Technicians (UCATT), schoolteachers and activists. In the UK, substantial research has gone into examining the dangers of embedded asbestos, and alternative framings now argue that domestic exposures to all types of asbestos are hazardous. Research published in early 2009 recognizes that people interact with the physical structures of their houses in ways not envisaged in HSE's management approach and asbestos regulations (Rake et al, 2009; Waldman and Williams, 2009). This alternative framing poses that in situ asbestos is very hard to control. It challenges the HSE's scientific understanding of risk and corresponding conclusion that in situ asbestos represents a miniscule risk, suggesting that these are based on the principle of non-disturbance of asbestos and on the assumption that buildings remain static. In reality, however, human beings live in and shape buildings through daily activities, maintenance, refurbishing and emergencies and, in the process, they often inadvertently disturb asbestos. Recently, there has been a constant stream of media reports highlighting this danger. There have also been several instances of schoolteachers contracting ARDs (Lees, 2005/2006; BBC News, 2009; McCulloch and Tweedale, 2008). In all these instances, one sees a similar trend: scientific experts have hastened to reassure the public that the asbestos is encapsulated and safe, while public concern remains high. The HSE, responding to queries about asbestos in schools has stated, 'it would be dangerous to remove asbestos sealed inside buildings' (BBC News, 2009). Meanwhile the National Union of Teachers (NUT) has declared asbestos in schools to be a 'ticking time bomb' (Keegan, 2009) and has requested that all asbestos be removed (BBC News, 2009). Paul Rowen, Rochdale MP until 2010, was a former schoolteacher who, for 5 of his 28 years teaching, was responsible for the maintenance of school buildings. He argued that 'councils are not doing enough' and activists have demanded that every UK school is regularly tested for asbestos.[17] This framing again emphasizes the environmental and social dimensions of asbestos and seeks to insert these priorities into mainstream policy in the UK. It thus draws on ambiguity, emphasizing how specific categories of people feel about their personal exposure to asbestos.

This framing posed by schoolteachers, unions and researchers, and arguing that the UK government is not doing enough to protect people, is challenged by another industry-driven alternative framing that argues the government does too much and overreacts to the dangers of asbestos. The UK's Asbestos Watchdog is an organization of research consultants, founded in 2003 to provide 'experienced and unbiased asbestos contractors' to advise the public on asbestos removal. Its website boasts of a 'panel of leading scientists in fibre toxicology, pathology, occupational lung diseases, and epidemiology'. Prominent members of the Asbestos Watchdog have established links with the asbestos industry.[18] Using arguments of bio-persistence, the Asbestos Watchdog argues that white asbestos undergoes substantial chemical and structural alteration when embedded in building products such as cement or plaster. This alteration is seen to significantly change the composition of chrysotile asbestos such that it no longer resembles asbestos as defined in the UK government's asbestos regulations (Bridle and Stone, n.d.). In 2009, the Asbestos Watchdog challenged a series of radio adverts run by the HSE. These were part of the HSE's 'Hidden Killer' campaign, designed to warn tradesmen, particularly construction and maintenance workers, about the dangers of in situ asbestos in old buildings. Several advertisements provided statistics[19] for the contraction of ARDs. The Asbestos Watchdog's John Bridle contested these figures – a process which the *Telegraph* headlines described as 'wilfully misleading' and sensationalist. Christopher Booker (2009), the journalist, summarized Bridle's claims, arguing that the HSE had failed to understand that white asbestos was substantially different to other kinds of asbestos and, as such, posed 'virtually zero' risk to people. The HSE's 'confusion' is said to be because powerful business interests have been allowed to promote the false assertion that chrysotile asbestos is hazardous. The HSE is thus accused by Booker of colluding with these business interests by 'putting out advertisements designed to panic the public into falling for the wiles either of the lawyers or of rapacious removal contractors' (Booker, 2009).[20] This alternative framing appears highly scientific in that it draws heavily on the language of risk. However, it uses different forums to academic science, publishing primarily in newspaper articles and not engaging in peer review. In posing questions about the proof of risk and challenging the claims that scientists put forward as evidence, it engages in what Wilson terms 'non-denial denial' tactic (2008, p67).

These alternative framings of asbestos and the related potential for disease are based on very different ways of knowing and do not always rely on scientific or biomedical calculations of risk. In general, UK alternative framings seek to complement or replace scientific and technological discussions about risk with social and environmental considerations. With the exception of the alternative framings provided by the Asbestos Watchdog, these framings emphasize emic and bodily ways of knowing that are, as is explored in earlier chapters, shaped by witnessing other ARD sufferers or by a person's internal and physical feelings after being diagnosed with an ARD. These forms of knowing create anxieties which are not resolved by formal calculations of risk because people, such as laggers and shipyard workers, know – with absolute certainty – that they too will eventually succumb to asbestos diseases. This is yet another instance of 'ambiguity' (Leach et al, 2007), or a situation in which the kinds of knowledge prioritized by activists, people exposed to asbestos and sufferers are very different to that prioritized by the expert-based approaches currently shaping UK policy. What the government labels as 'irrational anxiety' and misunderstandings of science is, to laggers and other asbestos sufferers, an absolute certainty. Similarly, the goals that people struggle for are justice, recognition of their experiences and protection for future generations. These are not the same as those dealt with under UK asbestos policy – which aims to minimize risk and to ensure a compromise between economic interests and workers' safety. UK asbestos activists are thus interested in more diverse understandings of governance and society, of multiple goals, and they pursue a range of pathways to achieve this (compare with Leach et al, 2007). Their alternative framings encapsulate workers' – and others' – sense of betrayal, not only from employers who failed to warn them about the dangers of asbestos and for which they paid with their health, but also because they feel 'cheated by an inequitable social and political system' (Tweedale, 2000, p288).

Scientific proof for asbestos risks: Alternative framings for India

Indian activists' alternative framings also stress the inherent dangers from asbestos and, as is evident in Chapter 4, reject the Indian government's science that 'proves' people exposed to white asbestos are not at risk. However, as argued in Chapter 4, it has been very hard for Indians to mobilize against asbestos. This stems from the success of externalizing the costs of asbestos manufacture onto disempowered and invisible workers (Kamat, 2008), the failure of the NIOH to register and report asbestos diseases (Dutta, 2008a) and from the narrow definition and formal recognition of ARDs as asbestosis (Murlidhar and Kanhere, 2005). ARDs are, as Dutta argues, 'hidden in India. It does not sing on trains or beg on the streets. The disease is misdiagnosed, underreported and forgotten' (cited in Daubs, 2008). Using the professional skills of its campaigners, the Ban Asbestos Network India (BANI) has recently focused on a scientific challenge to the government's position that asbestos is safe. This has involved encouraging unions to initiate health checks – a process which led one worker to comment that he had only became aware of asbestos dangers after 31 years of working with it (Kamat, 2008; Mohite, 2008; Rahman, 2008). Activists challenging the official narrative that asbestos is safe and the accompanying under-diagnosis of disease have sought to do so on the same terms as the government. They have focused on scientific research, rather than citizen science, aiming to amass the disparate recording of disease in India (Manavar and Patel, 2004; Krishna, 2006b; Daubs, 2008)[21] and, in so doing, to establish conclusive scientific proof that asbestos is dangerous. So, for example, in 2007, the Occupational Health and Safety Centre in Mumbai conducted medical inspections outside an asbestos factory gate and demonstrated that 23 per cent of the workers tested suffered from asbestosis (Gaitonde and Dutta, 2008). Similarly, in 2008, a 'Right To Information' submission forced the Tata Memorial Hospital to provide the first official evidence of mesothelioma. It reported that 107 cases of mesothelioma had been diagnosed between 1985 and 2005 (Dutta, 2008b). Activists have also highlighted instances where the government itself has recognized the dangers of asbestos. Zaidi, for example, begins an article as follows: 'The magic is wearing off. Now that asbestos has been officially acknowledged in India as the killer it has long been accused of being, the "magic mineral" is losing its popularity' (2006, p100). This 'official acknowledgement' of the dangers of asbestos stems from a Supreme Court Committee of Technical Experts that, in investigating the dangers of shipbreaking, found that 16 per cent of workers suffered from asbestosis. These alternative positions have thus focused on scientific knowledge of probabilities, trying to use the same language as that of the Indian government. Ultimately, however, these framings have not been able to challenge the dominant approach despite revealing internal contradictions, picking up on subtexts produced by the government itself, and seeking to facilitate the production of scientific evidence about risk.

Anti-asbestos activism has also revealed the misuse of science, questioning who funds research and interrogating scientists' neutrality (Nath, 2000; Ramanathan and Subramanian, 2001). Thus, in addition to exploring scientific, epidemiological debates regarding fibre size, bio-persistence and risk, activist research has examined political and economic considerations for continuing asbestos production (Nath, 2000; PRIA, 2004; Murlidhar and Kanhere, 2005; Joshi et al, 2006). Because workers are largely invisible, alternative framings have sought to find and mobilize widely dispersed workers, to encourage medical experts to recognize and look for occupational diseases and to persuade the state and the asbestos industry that 'isolated' incidences of ARDs are part of a widespread and under-recognized problem.

Alternative framings also draw on broader Asian framings in order to pressurize the Indian government. Since 2000, when six Asian delegates (including one from India) attended a Global Asbestos Congress in Osasco, São Paulo, there has been a steady growth in Asian awareness of asbestos hazards. Mobilization processes have matured through large-scale, international events that directed attention onto asbestos issues: in 2004 a Global Asbestos Congress was held in Tokyo; in 2006, the first Asian Asbestos Conference exclusively focused on asbestos in Asia; and, in 2009, a second Asian Asbestos Conference was held in Hong Kong. Attended by 'large contingents from India, Japan, Korea and Indonesia together with delegates from other Asian countries, global ban asbestos campaigners, medical experts, legal professionals, trade unionists, technicians and academics', the 2009 conference aimed explicitly to 'strengthen the grass-roots ban asbestos movement in Asia' (Kazan-Allen, 2009a, pp2, 3). In addition to the creation of the Asian Ban Asbestos Network (A-BAN), the international support at the conference facilitated the sharing of ideas and experiences and the linking of new grassroots participants with veteran campaigners from elsewhere in the world. The participants sought, throughout the conference, to emphasize two particular framings of asbestos, namely the voice of the victims and the collation of scientific evidence that proves the risk of asbestos exposure. Widowed women from China thus described how their partners had been incapacitated by ARDs (Kazan-Allen, 2009b), while experienced campaigners pointed out the significance of hearing victims' stories. For example, when Japanese widows reached out to other asbestos victims to share their suffering and loneliness, this led to the formation of victim associations and a national anti-asbestos campaign (Kazan-Allen, 2009b). As with many other alternative framings, this emphasis on ambiguity – drawing on individual stories rather than predictions of risk – can be seen as fitting within broader notions of incertitude and emphasizing the social dimensions of asbestos-related issues.

Many other presenters at the Asian Asbestos Conference stressed the collation of scientific evidence, emphasizing the lack of available information in Asian countries and pointing out that the lack of national statistics on incidences of ARDs facilitated Asian governments' lack of engagement.[22] Asian scientists presented research results which demonstrated high incidences of ARDs in Asia, with specific epidemiological and medical evidence coming from China (Wang, 2009); India (Malavadkar et al, 2009); Korea (Ahn, 2009); Japan (Hisanaga et al, 2009); and Indonesia (Kang et al, 2009). Linked to this scientific approach, unions emphasized the more political dimensions, showing how the Japanese government – despite banning asbestos – continued to under-recognize and under-compensate pleural plaques and other occupational diseases (Iida, 2009) and how Indian shipbreakers faced stiff opposition to their attempts to unionize (Sarde, 2009). While this framing fits within the scientific approach of risk, disseminating information through established scientists and academic forums, it also incorporates an element of incertitude, in that it highlights ambiguity in the way in which formal policy works and emphasizes how other factors – such as politics – can undermine a policy approach focused exclusively on scientific and technical conceptualizations of risk.

Alternative framings at the 2009 Asian Asbestos Conference thus emphasized victims' pain and suffering as a means of facilitating grassroots mobilization and produced scientific evidence of ARDs in many Asian countries. The conference concluded with a 'Hong Kong Declaration Towards a Complete Ban on all forms of Asbestos' and asbestos processing. This declaration recommended the implementation of policies to encourage safer technologies while ensuring the socially just phasing out of asbestos; sought greater administrative recognition and medical assistance for people suffering from ARDs; argued that companies exporting asbestos to southern or newly industrializing countries should be held criminally liable; and highlighted the need for international agencies, national governments and regional associations to be more closely connected with grassroots organizations and victims. These alternative understandings of asbestos – with their emphasis on broadening scientific risk by adding an awareness of ambiguity and highlighting social dimensions – were directly targeted at governments such as India's. During the conference, Indian anti-asbestos organizers discussed their difficulties in identifying and organizing asbestos victims, of getting asbestos conditions medically diagnosed, of linking compensation to occupational histories and of actually securing compensation (Dutta, 2009). Among the outcomes of the conference were two International Ban Asbestos Secretariat (IBAS) 2009 Grassroots Bursaries for India, which would allow the continued documentation of asbestos contamination in Gujarat, and would provide support to victim groups who have experienced asbestos mining and industry in the Udaipur district. Frustrated by the dominance of official approaches within India, Indian activists have drawn inspiration and support from regional debates happening beyond the national border and shaped their alternative framings in response. This pattern of seeking to influence national debates by drawing on regional and neighbouring countries' experience is echoed in the following section, which explores alternative understandings in relation to South African asbestos debates.

Regional development and poverty: Alternative framings for South Africa

South Africa banned all asbestos production in 2008, as mentioned above, with strong penalties for anyone processing, manufacturing, importing or exporting asbestos-containing material. In so doing, the South African government has positioned itself as a 'government of the people'. This involved a complicated transition – conducted in the dying days of the asbestos industry – in which unions and industrial manufacturers of asbestos-containing products came to accept the scientific definition that all asbestos is carcinogenic. This means that opposition to the South African official framing of asbestos comes largely from South Africa's only asbestos-mining neighbour, Zimbabwe, which argues that 'environmentalism and occupational health are luxuries that Africa cannot afford' (McCulloch and Tweedale, 2008, pp256–257). After the South African National Asbestos Summit of 1998 (described above), Zimbabwe's asbestos industry benefited as South Africa introduced new health and safety regulations.[23] This led South African companies to import cheaper Zimbabwean chrysotile and asbestos-cement products (Blandy, 2006). By 2006, when the South African Department of Environmental Affairs and Tourism started gearing up for a complete ban on asbestos, Zimbabwe had a vested interest in ensuring the continued use of asbestos in South Africa. An estimated 80,000 Zimbabweans worked in asbestos mining or manufacture (Makoshori, 2008), South Africa was Zimbabwe's largest importer of asbestos, and Zimbabwe needed to transport asbestos through South African ports to access foreign markets.[24] Zimbabwe thus sought to counter the proposed banning of asbestos. A high-level government-to-government initiative – the Chrysotile Joint Technical Task Force Team – was established as intense negotiations occurred between the two countries. The Task Force met in Harare in October 2006, and again in Cape Town in June

2007. Zimbabwe's delegates were anxious to structure discussions around the safety of chrysotile asbestos, however, the South African delegation wished to avoid rehashing the debates around the health risks of asbestos. Instead, South African representatives argued that the Department of Environmental Affairs and Tourism had no option but to implement a total ban of asbestos according to the Cabinet Memo that emerged out of the National Asbestos Summit of 1998. Nonetheless, South African delegates recognized Zimbabwe's right to continue to mine and manufacture asbestos.

Alternative framings presented by Zimbabwean delegates were thus twofold: first, that sufficient scientific evidence existed to prove that white asbestos was safe while more research was needed to ensure that asbestos substitutes were risk-free, and second, that economic considerations were obscuring the objectivity of the science. Echoing dominant Indian perspectives, science was invoked to support the framing of Zimbabwean chrysotile (white) asbestos as safe, because it had low bio-persistence and was being used according to International Labour Organization (ILO) Convention 162 (Gibson, 2007b).[25] Zimbabwean medical personnel argued that their examination of 70,000 medical records had not identified a single mesothelioma sufferer and that workers exposed for 25–40 years remained healthy. The South African burden of ARDs stemmed, in their view, from the mining of crocidolite and amosite (blue and brown asbestos). Zimbabwean chrysotile could therefore be safely manufactured (Blandy, 2006).[26] Furthermore, the Zimbabwean delegation drew on scientific notions of risk to argue that there was no 'grounded scientific evidence' to prove that asbestos replacements were risk-free (Matambanadzo, 2007a). More research was therefore needed before introducing a ban on asbestos in order to prevent repeating the debate when, in 20 years' time, the latency period of asbestos substitutes emerged. This Zimbabwean narrative around asbestos replicates the 'non-denial denials' made by the UK's Asbestos Watchdog. Despite being an alternative framing, it uses the language of science and risk but fails to engage in conventional academic forums of experimentation, peer review and publication.

Zimbabwean delegates also examined asbestos mining in economic terms. They argued that South African asbestos manufacturers were supporting the asbestos ban to 'have the market [for non-asbestos alternatives] for themselves' and that their campaign was motivated by economic greed. Zimbabwean companies needed an opportunity to develop alternatives and, for this reason, the ban should be postponed. There were, in addition, potential job losses for Zimbabwean miners (Gibson, 2007b; Blandy, 2006). Banning asbestos, it was argued, would 'plunge about 100,000 people into poverty' (Gqubule, 2006). On this issue of economic imperatives, the Zimbabwean lobby had some support from South African small businesses. These small traders argued that big business – such as Everite, described above – were advocating for the asbestos ban in order to block competition from black businesses. As these businesses had not had the same research opportunities, they were constrained by Everite's monopoly on alternatives (Blandy, 2006).[27] This framing can be seen as a form of ambiguity in that it stresses the alternative values associated with employment and money, prioritizing these above workers' health. In so doing, the social dimensions, articulated through workers' voices or victims' personal accounts, are erased.

Despite South Africa's absolute confidence that all asbestos was hazardous and that the ban on asbestos would be introduced, the Zimbabwean media stressed that the Task Force had extended the deadline of the draft Ban on Asbestos to allow for more consultation and to ensure that Zimbabwe could make further submissions. This implied that there was still a possibility that asbestos would not be banned in South Africa. It also quoted directly from the joint statement produced by the Task Force: 'The joint technical taskforce recognises and respects that Zimbabwe still produces, uses and exports and handles chrysotile asbestos-related products in accordance with the provisions of their own laws and international conventions such as the ILO conventions 162 and 155' (Matambanadzo, 2007b). It is clear from the discussions at the Task Force that the South African delegation did not agree with this interpretation, arguing that these international conventions had been superseded by later ILO and World Bank pronouncements – but that this was, in effect, a face-saving gesture for Zimbabwe. The process of establishing a task force and the associated meetings provided a framework for moving forward. It was agreed that South Africa would investigate what kinds of regulation were appropriate for the transport of Zimbabwean asbestos through South Africa and that Zimbabwe would initiate inquiries into alternative fibre technology.

Alternative framings and national contexts

The alternative framings discussed above and summarized in Table 6.2 below demonstrate a wide variety of, sometimes contradictory, perspectives. On the whole, alternative framings perpetrated by marginalized victims of ARDs, by environmentalists, working-class people, trade unionists and some scientists seek to highlight incertitude by emphasizing different values, introducing ambiguity and projecting social and environmental concerns. They also illustrate the importance of context, of national policy processes and of diverse actors' positioning. Some alternative framings are, however, heavily invested in 'expert' science and in the 'rational' production of scientific calculations of risk, such as those propagated by Zimbabwean officials, industry and by conservative (and industry-linked) critics in the UK.

Alternative framings aim primarily to influence national policy processes by asserting 'a new set of interests into research that represents marginalised groups'

(McCormick, 2009, p5), or – as is clear from Table 6.2 below – by stressing incertitude. Activists are thus seeking to reframe scientific objectivity and to challenge official science. Their alternative framings do this by introducing stronger social and environmental considerations, emphasizing the personal experiences of asbestos victims, in terms of anxiety, stress and awareness, and showing the need to factor socio-economic context and environmental processes into policy, regulation and assessments of risk.

Governing through science and technology

Jasanoff (2005) argues that technical advisory councils, court proceedings, regulatory assessments, scientific controversies and even the transient websites developed by environmental groups and multinational corporations provide illustrations of the practice of democracy. Thus the juxtaposing of dominant and alternative framing of asbestos issues in any one country reflects a combination of political culture, national context and national styles of decision-making. In addition, and as is evident in the above discussion, the forms of expertise that are recognized in particular countries are themselves culturally shaped and vary from country to country (compare with Jasanoff, 2003). It is thus possible to distinguish stylized ways of 'doing politics' in each country. Asbestos framings in the UK demonstrate a tension between a 'knowledge society' – in which knowledge has become the primary resource and state policies prioritize scientific and technical expertise (Jasanoff, 2005, p4) – and alternative ways of understanding problems. Dominant framings of asbestos are thus focused on the production and contestation of medical and legal knowledge, rather than asking broader questions about how people experience disease. A top-down, expert-informed approach is adopted in the UK that – in casting expert knowledge as unproblematic – obscures the relationship between knowledge and power. Citizen participation is nonetheless encouraged as a means of strengthening state-society relations, with demonstrations in parliament, activists lobbying their MPs, or through government inquiries and representations. Activists are thus forced to engage with science and the language of risk regulation in order to challenge dominant framings. The Dagenham laggers, for instance, are beginning to produce their own lists of diseased people and, from these lists, to develop their own statistics. In March 2010, they recorded that 60 members of their GMB branch had been diagnosed with ARDs since 1998. Of these, 32 men were now dead. Although this provides one form of proof to them, ultimately, it is the scientific opinions of government-recognized experts and their dominant framings which hold sway.

Table 6.2 *Alternative framings and primary actors in the UK, South Africa and India*

	Alternative Framings	Primary Actors
UK	Heightened awareness that in situ asbestos might be lethal, with the HSE providing inadequate legislative protection.	Trade unions, especially UCATT, school teachers and activists, select MPs, some social science researchers.
	Pleural plaques are not benign and symptomless but are an occupational disease and precursor of other ARDs; brings in understandings of history, particularly of industrialization and class issues.	People involved in lagging, shipbuilding and railways – industrial working class, Scottish Parliament, asbestos sufferers, victim support groups, lawyers and medical specialists experienced in representing sufferers in asbestos cases.
	The HSE overreacts and over-emphasizes the dangers of asbestos.	Industry-linked research consultancies such as the Asbestos Watchdog.
	A reframing of anxiety.	Activists, people exposed to asbestos (laggers, shipbuilders, people working on the railways) and ARD sufferers.
South Africa	Environmentalism and occupational health are luxuries that Africa cannot afford.	Zimbabwe's asbestos industry and medical specialists employed in that industry, small South African black empowerment businesses, Zimbabwe government officials.
	Sufficient scientific evidence exists to prove that white asbestos is safe, and there is no scientific evidence that asbestos substitutes are safe,	Zimbabwe's asbestos industry and medical specialists employed in that industry, Zimbabwe government officials.
India	There is conclusive scientific proof that chrysotile asbestos is dangerous.	Scientists, journalists acting under the auspices of BANI, other Asian actors including medical scientists and doctors from China, India, Korea, Japan and Indonesia, government officials from Japan, shipbreakers from Bangladesh and India international anti-asbestos activists such as IBAS.
	There are political and economic vested interests that perpetuate asbestos production.	Some trade unions, BANI activists.
	Asia is listening to victims' voices and responding appropriately.	Widowed women from Japan and China.

India too is characterized as a 'knowledge society' (Shah, 2009; Stirling, 2009) and the reliance on scientific, technical paradigms in asbestos policy is only too evident. This approach relies on an extremely narrow definition of risk where scientific parameters are used in unscientific ways. A top-down, risk-orientated governmental approach determines both who the experts are and how science is used. Activists challenging this dominant framing of science, risk and disease have sought to introduce social and environmental considerations, have challenged the epistemology underlying the science and have facilitated participatory engagements with government, but with little success. Risk regulation in knowledge societies, such as the UK and India, have inquired whether asbestos – in its various incarnations – is 'safe'. Ideas about what constitutes 'safe' have been narrowly defined by experts and governments (Stirling, 2009). These dominant framings thus fail to examine ambiguity (between different kinds of disease for instance), and ignore comparisons of who is harmed and how (whether laggers are more affected than electricians, shipbuilders or shipbreakers). The entire focus of UK and Indian government policy asks 'what is the chance of something happening' and these chances are assessed by these governments through different calculations and perceptions of risk.

In contrast to the UK and India, since the 1990s South Africa has approached asbestos debates with a degree of suspicion, strongly influenced by the manner in which science had been dominated by white elite and by medical scientists' complicity during the apartheid era (compare with Wilson, 2008). A strong desire for reconstruction and a concern for the poor, coupled with a need to give the poor 'ownership' of their development, has fed into asbestos negotiations. The result is a broader scope for non-expert voices to be heard in policy and a broadening out of knowledge to encompass disadvantaged and marginalized perspectives. Such knowledge is far more contingent, diverse, ambiguous and relational. While this approach has clearly not solved asbestos problems, it has resulted in citizens having a greater sense of ownership and control over the problem, in fewer South African alternative framings and in greater awareness – across all sectors of society – of the dangers of asbestos. The approach has thus not been to ask 'what is the chance of something happening?', but rather 'how can we best protect against something happening or, if something has already happened, how can we ameliorate the experience?'

Conclusion

Science, as McCormick argues, is positioned at the intersection between corporate, political and activist interests; 'The multiplication of and contest over expert knowledge and technology are pathways through which these multiple interests intersect' (2009, p27). Dominant and alternative framings are thus contesting interpretations of science which emphasize different concerns and values. Both UK and Indian dominant framings of asbestos focus on medical and technological aspects with the exclusion of social and environmental issues. Thus, the authority of 'particular views of systems and their goals to the exclusion of others is not mere chance; it also reflects politics and power' (Leach et al, 2007, p8). McCormick suggests that there are three main reasons for this. First, in general, biomedical and scientific paradigms are seen as more plausible and less contentious than environmental understandings. Second, research of this nature is easier to undertake (in terms of research design, testing causality and available funding) than environmental health research. And third, unlike environmental health research, biomedical research on disease causation avoids questions of responsibility, accountability and regulation (or the failure thereof). It does this by focusing exclusively on the individual at the expense of broader environmental, social and political dynamics (McCormick 2009, p76). The inclusion of social and environmental factors radically reshapes – as the South African example demonstrates – the dominant scientific and technological framing. Nonetheless, framings are not always rigid. Although dominant framings may have a propensity to overlook conditions of incertitude (Scoones et al, 2007), the inclusion of notions of ambiguity can, as the South African example shows, be particularly powerful. The comparison between South Africa, India and the UK also demonstrates how framings may be both alternative and dominant depending on context. Comparison also reveals the degree to which alternative framings seek to emulate – perhaps not always successfully – and undermine the scientific strategies of dominant framings, by drawing on precisely these strategies.

The case of asbestos demonstrates, as McCulloch and Tweedale argue, that 'science is never sufficient to solve occupational health problems' (2008, p275). There is no shortage of information about asbestos and its potential health hazards. Indeed, as this chapter demonstrates, all discussions about incertitude focus on ambiguity, rather than on uncertainty or ignorance. This suggests that there is no shortage of statistical quantification and very little sense of outcomes that are still unknown (with the exception of some alternative framings that argue that the effects of asbestos substitutes are unknown). In all three countries, alternative framings have been – and continue to be – an active process of contestation regarding what kinds of considerations should be incorporated when dealing with asbestos-related issues: to what extent should economics be considered, are workers' livelihoods more important than their health, what status should victims' voices and personal accounts be given, how significant are environmental considerations and to whom should accountability be accorded?

Determining what kinds of facts should count is a process which is normative, rather than scientific, and is, ultimately, about value judgements. Questions regarding environmental and social sustainability and social justice, as explored in the following chapter, are thus woven into dominant and alternative framings of asbestos.

Notes

1   For instance, the UK Health and Safety Executive (HSE) estimates that 1.5 million workplaces contain in situ asbestos (Waldman and Williams, 2009).
2   In addition, different sampling techniques, different times of day and sampling at different heights often produce different results (Chesson et al, 1990; see also Tweedale, 2000).
3   In 2007, the World Social Security Forum recognized that special measures are required to deal with in situ asbestos. It argued that the management of this risk presented many financial and technical difficulties. Overall, it 'may be better to maintain the asbestos in place for as long as possible'. If removal is necessary, the forum emphasized the need for wet removal which retains the 'integrity of the substance' rather than destruction and the need for careful and safe disposal which avoids possible reuse (Leprince et al, 2007; see also Commins, 1991).
4   Health risks are influenced by a wide range of factors, including: length of exposure; type of asbestos fibre; length, diameter and distribution of fibres; age; sex; smoking habits or exposure to passive smoking; other occupational or domestic exposures to toxins; health status; conditions of exposure and so forth (Bignon, 1989). In addition, different mechanisms for assessing airborne fibre concentrations will produce different results and affect estimates of health risks. Hughes and Weill produced the following estimates for annual risk of death from asbestos exposure per million people (as contrasted to other death risks from daily activities): 'studying in a school sprayed with asbestos, 0.25; cycling to school from 10 to 14 years of age, 15; inhalation or ingestion of foreign bodies, 15; playing football at school, 10; chronic smoking, 1200; passive smoking for two months, 1' (cited in Bignon, 1989, p22).
5   In determining whether asbestos-containing materials are 'high' or 'low' risk, and whether the degree of damage/decay creates a risk of exposure, the HSE relies on scientific evidence. It emphasizes the values of different types of surveys, the need to monitor fibre levels and – in the event of asbestos removal – to confirm these levels through a certificate of reoccupation.
6   In the case of extensive calcification – or very large pleural plaques – the report recognized that these might inhibit breathing, but experts consulted by the IIAC concurred that this affected less than 1 per cent of pleural plaque sufferers.
7   Including simplification that removes all the detail from the story, fabrication that rearranges the remaining elements, demotion of the claimant's motive, characteri-zation that reframes the question of legal responsibility and blameworthiness into

moral questions, inflation of the actual amounts of compensation, and a process of normalization which, through selective profiling, makes exceptional cases appear routine (De Saulles, 2006c).

8   A fibre, thought to be capable of causing ARDs, is scientifically defined as being greater than 5 micrometres, having a width that is less than 3 micrometres and an aspect ratio of more than 3:1 (Burdett et al, 1989). Asbestos bodies smaller than fibres are generally termed particles.

9   In this case, BANI mobilized indirectly through Ahmedabad-based NGOs to influence local opinion. Surrounding communities were made aware of the dangers of asbestos and wrote letters protesting against the factory. Representatives from these communities demonstrated at the EPH, holding placards that stressed the dangers of asbestos. In addition, scientific-based NGOs scrutinized the proposals and stressed, in the EPH, the failure to explain how the regulations would be met. For instance, the Safety, Health and Environment Association questioned the provisions for measuring airborne asbestos fibres, the types of occupational diseases possible and the safety handling procedures.

10   For example, when South Africa became a signatory to the UN Environmental Council Agreement in 2001, asbestos could no longer be used in car manufacture. Companies made considerable investments into researching possible alternatives.

11   Companies still processing asbestos were allowed a three-month grace period. The regulations also set out the circumstances in which – under particular, strictly controlled conditions – exceptions might be granted.

12   Everite stopped asbestos production in 2002 – well in advance of South African national requirements to do so.

13   Intensive surveying was designed in order to determine accurate rates of contamination, which facilitates improved extrapolation to other areas and allows for more precise calculations of the total cost of asbestos rehabilitation and remediation in South Africa.

14   In particular, the Limpopo, Mpumalanga and North West Provinces.

15   Established in 2003, the Asbestos Relief Trust (ART) is the outcome of an out of court settlement between claimants suffering from asbestos conditions and the following mining companies: Griqualand Exploration and Finance Company, General Mining Corporation, Msauli, African Chrysotile Limited and Hanova Mining Holdings. The purpose of ART is to compensate employees of these companies who are currently, or may in the future, be diagnosed with ARDs. Because the General Mining Corporation contributed to the Cape plc settlement described in Chapter 3, employees who were claimants of the Cape plc case are excluded from this trust. ART also administers the Kgalagadi Relief Trust; a fund established in 2006 on behalf of the former Swiss Eternit Group to ensure that former employees receive financial assistance if/when they contract asbestos diseases. The fund covers workers of the Kuruman Cape Blue and Danielskuil Cape Blue asbestos mines between 1952 and 1981, dependents and children of employees and people living in the vicinity of these mines.

16 A Private Members' Bill is a means by which MPs can seek to bring about a change in the law. Although, in practice, very few Private Members' Bills do actually bring about a change in the law, they are an effective means of raising publicity and alerting other MPs and Lords to particular issues.

17 www.asbestosexposureschools.co.uk/npaper%20articles.htm and www. theyworkforyou.com/whall/?id=2009-03-25c.124.0, accessed 11 November 2009.

18 The Asbestos Watchdog has many similarities with the Chrysotile Institute described in Chapter 4.

19 The HSE advertisements gave the following statistics for asbestos-related deaths: 6 joiners, 6 electricians, 3 plumbers and 20 tradesmen affected weekly. The Advertising Standards Agency concluded that these were inflated by approximately 10 per cent and that, realistically, fewer electricians (5 instead of 6) and tradesmen (18 rather than 20) were affected weekly. It also argued that the HSE should have clarified that these figures were based on estimates and did not represent actual deaths (Wilson, 2009).

20 While the Asbestos Watchdog's alternative framing presents 'conclusive' science to prove that asbestos embedded in cement and other building products is safe, these claims are not widely accepted and have been publicly challenged. In a series of blogs, media commentaries and books, critiques have pointed out the similarities between these arguments and those of the asbestos industry and the close connections between the Asbestos Watchdog's experts and the asbestos industry (Wilson, 2008). They have also challenged the Asbestos Watchdog's scientists' qualifications and expertise (Monbiot, 2008).

21 The Occupational Health and Safety Centre in Mumbai diagnosed 41 out of 99 asbestos factory workers with asbestosis; the Tata Memorial Hospital in Mumbai recorded 107 cases of mesothelioma between 1985 and 2005 (Daubs, 2008); the Industrial Toxicology Research Centre, India, recorded nine cases of asbestosis at the Ahmedabad power plant (Manavar and Patel, 2004). The NIOH estimated, in August 2006, that one in six workers — 16 per cent of the workforce that handles asbestos — could be suffering from asbestosis (Krishna, 2006a).

22 Only two Asian countries have ratified the International Labour Organization (ILO) Asbestos Convention 162, while six have banned asbestos. The Asbestos Convention is part of the ILO's desire to bring about a global ban on asbestos, which it now recognizes as the one of the world's *greatest* industrial killers. In June 2006, at the 95th Annual Conference in Geneva, the ILO resolved that the eradication of all future asbestos use, in conjunction with proper identification and management of in situ asbestos, was the most appropriate way of protecting workers' exposure. It also stressed that the Asbestos Convention 162 of 1986 should not provide a justification for the ongoing utilization of asbestos (ILO, 2006).

23 Although most South African amosite and crocidolite mines had closed in the 1980s, some chrysotile mining continued until the introduction of these Asbestos Regulations in 2001 and the ban in 2005 (Kgalamono et al, 2005).

24 The 180,000 tons produced annually was exported to South Africa and Brazil before these countries introduced bans in the late 2000s, as well as to India, Iran, Dubai and China (Makoshori, 2008).

25   Although the ILO had revised its position in June 2006 and recommended a total ban on asbestos (see note 22).

26   Invitees to the Task Force, who represented South African asbestos manufacturers, pointed out that, in the past, South African companies had mined the same mineralogical seam and many of these workers were now being diagnosed with ARDs and, in so doing, echoed the dominant framings produced by the South African government (see also McCulloch, 2003). In addition, chrysotile is seldom mined in its pure form and is often contaminated by amosite and crocidolite fibres.

27   Everite begun to explore asbestos substitutes in 1985, and, in 2001, with the asbestos ban looming, invested R140 million into researching new technologies. It examined more than 300 alternative fibres before settling on Nutec – a combination of Portland cement and cellulose organic fibre (as found in paper and cardboard), used in combination with polyvinyl alcohol (PVA), condensed silica fume and aluminium trihydrate. Although some of these chemicals can damage people's health when used in large quantities, Everite argues that these are usually thought to be safe (Gibson, 2007a).

Chapter 7

Conclusion: Diseased Identities and Social Justice

Science is essential for democracy but is not in itself democratic.
(McCormick, 2009, p4)

The world has changed radically since the heyday of asbestos mining and asbestos production. These changes have been brought about, not because science has shown asbestos to be carcinogenic, but rather because people have mobilized and challenged governments to implement changes. McCulloch, Tweedale, Jasanoff, Castleman and many other authors have examined the historical trajectory of asbestos companies, medical science and regulatory policy processes. As their work has shown, and the chapters in this book underscore, science and technology are not neutral and do not always 'do' what is right for vulnerable people. The history of asbestos mining, production and regulation policy conclusively demonstrates that one cannot expect industry to look after its workers, either by acknowledging the negative effects of its activities or by implementing and following appropriate regulation, and one cannot assume that policy adequately meets people's health needs. As I write this conclusion, anti-asbestos movements, trade unions and ordinary people continue to try and effect change. In South Africa, they are grappling with the enormity of asbestos pollution and the scale of disease while struggling with under-resourced state institutions; the asbestos relief trusts are introducing new and innovative ways of facilitating employment verification and asbestos claims; and spokespersons from South Africa are engaging in the global ban asbestos campaign, contributing insights and lessons from the South African experience. In the UK, MPs and government lawyers are meeting to re-examine the question of compensation for pleural plaques, yet another government consultation on occupational disease is taking place and newspaper articles yet again debate how dangerous white asbestos is. In India, the presence of the Quebec Premier at the Delhi Sustainable Development Summit is reviving protests over asbestos and Canada's exportation of this product to India, while more and more cases of mesothelioma and other asbestos-related diseases (ARDs) are coming to light.

This book has sought to explore the role of science, the use of law and the criteria for medical conditions associated with asbestos from the perspectives of the powerless: those who are diagnosed with ARDs; who live in asbestos-polluted environments; or who mobilize around asbestos issues of recognition, compensation, retribution and rehabilitation. The chapters and stories produced in this book have shown that policy, regulation and compensation – in whatever format – seldom conform to people's experiences and ultimately have little traction for those seeking to live with the diagnosis of an ARD, a history of asbestos exposure and/or environmental pollution. In so doing, this book has not only examined how scientific knowledge is converted into power, it has also facilitated the exploration of processes by which 'power converts interests and power into science' (Harding, 1998, p51, cited in McCormick, 2009, p5). It has demonstrated how, with regards to asbestos, science has often been used to slow down – rather than speed up – regulation. As Tweedale (2000) argued in his examination of asbestos production and regulation in the UK, the political and social safeguards that enable knowledge to be used for public benefit were lacking. Effective regulation thus requires political will rather than more science and more sophisticated technical evidence. Science, technology and risk are not neutrally inserted into processes of governance. But science and technology are nonetheless critical to democracy because of the manner in which 'important aspects of political behaviour and action cluster around the ways in which knowledge is generated, disputed and used to underwrite collective decisions' (Jasanoff, 2005, p6). The practice of democracy occurs, as Jasanoff points out, not only around elections and voting, but through science and technology policies, in advisory committees, legal processes, risk assessments and scientific controversies.

The detailed examination of South Africa, India and the UK's asbestos issues shows how knowledge about asbestos and ARDs undergirds the processes of governance and governmentality. Medical definitions of the different kinds of ARDs do not facilitate improved treatment, but do inform patients and victims how they should behave and what kinds of bodily sensations they should be experiencing. Thus we see in Chapter 2 that UK pleural plaque patients are instructed that they should, at the most, experience only slight breathlessness. Otherwise, they are fine to continue with their lives as normal. This knowledge, in turn, feeds into legal debate and discourse and, as is evident in Chapter 6, becomes the basis on which legal framings of deserving (or undeserving) citizenship get made. In India, the failure to recognize the dangers associated with asbestos and hence to diagnose or label patients' experiences also shapes their knowledge and behaviour as citizens. In effect, it encourages a passivity which, in turn, hampers anti-asbestos mobilization while perpetuating workers' and others' faith in economic growth – underpinned by scientifically tested processes of risk assessment, regulation and monitoring – as a development trajectory.

In stark contrast to these two examples and, as shown in Chapter 3, in South Africa the widespread acknowledgement of people's illness and the corresponding systems of compensation encourage people to act in political forums and to engage with government and other stakeholders regarding environmental, health and other issues. While acting as participatory, enfranchised and knowledgeable citizens in this one rural town context, as explored in Chapter 5, the widespread awareness of illness and compensation also leads to South Africans consciously and continually seeking official diagnoses. Expert knowledges and scientific techniques encapsulated in legal frameworks and health provision thus inform people about how they should act – and feel – in relation to ARDs and what appropriate forms of citizenship should follow these behaviours and attitudes.

This 'education' through science, technology and law often leads individuals and collectivities to regulate their personal and bodily dispositions to fit with official framings and dominant political and economic objectives (Long, 1996; Moore, 1996). However, as the chapters in this book make clear, this is not always the case in relation to asbestos. Instead, a range of factors lead individuals and collectivities to challenge processes of governmentality and their 'education' into appropriate citizenship. Dominant framings, by and large, do not fit these people's experiences of everyday life. Dominant framings do not match the bodily sensations felt by asbestos sufferers – not least because these bodily sensations are not experienced in isolation from social and environmental factors. Dominant framings do not, with the possible exception again of South Africa, provide avenues for meaningful interpretation of ARDs.

In expressing their disagreement with dominant framings, individuals, organizations and social movements are also asserting their identities – that often take a particular form in relation to asbestos. These identities are about 'place', seen both geographically and in relation to broader societal structures and status. Thus, the laggers' masculine identities are a product of growing up in East London and of their economic and class marginalization. The South African asbestos sufferers studied are proletarianized workers suffering from unemployment, poor education, very limited economic possibilities and widespread environmental asbestos pollution. In India, sufferers of asbestos do not have an opportunity to develop specific identities, in part because their collective identity as workers is undermined by part-time, contract and temporary employment and in part because migration erodes the possibility of locating particular defined geographic spaces where asbestos problems are prevalent. Ban Asbestos Network India (BANI) activists, however, are both enabled and constrained by their urban, educated status and are able to take advantage not only of their professional identities, but also to mobilize across interconnected political 'spaces'. Finally, people's identities are shaped – ironically perhaps – by science itself. In both South Africa and the UK, despite the differing geopolitical contexts, the diagnosis of an ARD results in stereotypical gendered identities being asserted – women respond by locating themselves within the domestic arena and men by emphasizing their work as income earners in the public domain. There is an irony to this 'hardening' or reification of identities through science, namely that it works in contradiction to social mobilization, which is often seen to require more flexible identities and results in new identities. It may be for this reason that activists campaigning against asbestos issues seldom align with more prominent social mobilizations or go on to address broader campaigns for social or environmental justice.

In the three countries discussed in this book, the people and the movements involved are aiming, not to hinder the development of scientific knowledge, but to make it more accountable to the concerns of the poor, marginalized and disempowered. They have in common ambitions about constructing a future that is 'better', better for them as men and women, better for the environment and better for the nameless asbestos victims of the future. Asbestos mobilization can thus be seen as a way of challenging the science/knowledge/politics and power nexus while seeking to create new understandings of knowledge influenced by people's experiences, identities and concerns. At the beginning of this book, it was suggested that a considerable amount of thinking with and about science is still needed in order to understand the interfaces between science and society and to advance democratic possibilities. It was furthermore suggested that exploring asbestos mobilization will reveal new democratic possibilities and help identify pathways to greater sustainability and social justice. As discussed earlier, sustainability refers to the integration of 'specific qualities of human well-being, social equity and environmental integrity' (Leach et al, 2010, p5) into political processes through social mobilization, public deliberation and negotiation. It refers to the ways in which different people try, in a context of asbestos contamination, to articulate their values and seek to integrate these into government policy in ways that shape well-being, ensure social equity and sustain environmental integrity. How then, do these detailed accounts of South African, Indian and UK asbestos experience and mobilization illuminate the ways in which science and technology can be used to further broaden notions of sustainability and social justice? Chapter 6 discussed the dominant framings of asbestos often evident in government and policy circles and contrasted these with alternative framings. Linked to these various and often contradictory framings are different possibilities of action, or pathways, to social justice and sustainability. These pathways – both operating in practice and perhaps emergent or embedded within alternative framings – can be seen as knowledge collages constructed through different sets of 'useful' knowledge (Harding, 1996) or as 'alternative politics of sustainability' (Leach et al, 2010).

It is clear that, in all three countries examined, alternative narratives construct the problem in ways that differ from dominant framings. These alternative framings can be collectively grouped as follows: a concern for specific places or people (rural mining areas, industrialized sites) that experienced massive asbestos exposure and that still suffer the consequences; a sense of destruction of self through the experience of the disease; a fear of the future; revelations about gaps in regulation and the need for further protection; and a focus on victims' exploitation. Alternative understandings – while seeking to challenge dominant policy framings – cannot change policy or influence governments simply by their existence. In order to do so, they require 'traction' within policy processes in order to develop into alternative policy solutions.

In identifying the possibilities for traction, three pathways emerge as the most fruitful. The first comes from the South African example and is concerned with the voices and perspectives of the poor and marginalized. This pathway was particularly effective in this context because it exposed policy-makers to alternative realities in particularly visceral ways: showing them the abandoned asbestos dumps, presenting sick people before them to see their bodily deterioration and making them walk through economically depressed rural towns. This shaped government representatives and MPs perspectives, helping them to see differently and to experience poor people's day-to-day struggles in context. It also helped the Concerned People Against Asbestos (CPAA) to identify powerful people who supported their cause, to explore different ways of tackling local sustainability problems and to be engaged in local policy processes.

The second pathway to sustainability is highly interconnected with the first and concerns the engagement of activists and victims in science and technology. This is, in effect, a politicization of science. This involves a degree of serendipity and spontaneity, making the most of opportunities as they arise and – in conjunction with these – creating possibilities for new opportunities. In South Africa, politicization occurred simultaneously at both national and local levels, in part through serendipity and in part through the actions of the CPAA and other activists. Thus the political transition from the apartheid government to a democratic state occurred at roughly the same time that a range of social and environmental asbestos issues were being politicized in small rural towns. The former led to an awareness of how science had been used to benefit previously advantaged South Africans and to shore-up elite vested interests, while the latter helped people realize that asbestos-related illnesses were not natural phenomena, but stemmed from political and economic priorities that deliberately marginalized workers and rural town residents. The convergence of these two processes of politicization greatly advanced activists' agendas and led government authorities to seek new ways of engaging with science. These processes did not substantively change the nature of science, but did shape people's and policy-makers' perceptions of the need to tailor science to particularly vulnerable communities. This broadening out of policy processes is reflected in South Africa's dominant framings of asbestos, which seek to address concerns with job losses, environmental pollution, people's well-being and the science of asbestos simultaneously in order to find the most appropriate solutions. Thus, although the South African government has not been able to address the problem of asbestos embedded in building structures any more than the UK government, it has managed to bring together a far wider range of stakeholders and perspectives and to address the problem in a manner which allows greater possibilities for communities to consider their exposure to asbestos and provides individuals with greater prospects for financial redress.

The third – albeit still emergent – pathway to sustainability and social justice lies in the building of networks across political and geographic spaces. Although pursued by South African, UK and Indian activists to varying degrees, this is most successful in the South African case where networks and alliances criss-crossed local, national and international contexts. In the Indian example, anti-asbestos mobilization initially focused on building networks and creating 'spaces' through international ban asbestos organizations and strategically situated activists, and using these to challenge the Indian Government's use of science and technology to shore up asbestos-related economic interests. More recently, BANI has sought to build more south Asian political connections, tapping into an Asian science agenda to provide evidence of asbestos hazards. It has also, simultaneously, capitalized on its international linkages to help build new relationships, in particular with the 'missing' workers and to develop a new focus on rural contexts in India. This strategy – that has not yet led to changing policy around asbestos – is combined with the highlighting of government inconsistencies in the use of science and a strategy to politicize science. Still lacking from these networks are powerful actors positioned *within the government* who can act as brokers and convince others of the need to engage and reflect on different understandings of asbestos and its dangers. This is, nonetheless, an emergent pathway which 'has traction' because it is open to opportunities, contingencies and serendipity; because it seeks to bring together actors with different identities; because it will now build on the voices of the poor and will, in the coming years, put increasing emphasis on people's experience of ARDs; and, ultimately, because it is still in the process of campaigning for a ban.

Indeed, although activists and collectivities in all three countries have combined diverse strategies in order to change policy for the 'better', the UK example can be seen to have the least traction in policy and may thus be the least successful in addressing questions of sustainability and social justice. This is, in part, because a ban on asbestos has already been implemented. This creates a widespread perception that the problem has already been dealt with and no longer requires sustained attention. This perception is, in turn, reinforced through the use of technical scientific parameters. The ban also has the effect of marginalizing UK laggers from the international ban asbestos activities and activist networks. In part, and in contrast to the pathways identified above where change moved from more politically conservative to greater inclusivity and an awareness of the experiences of the poor, this may also be due to the political changes in the UK: a more conservative government is unlikely to result in a broadening out of science and technology. Increased political uncertainty thus brings with it increased resistance and a shoring-up of political, scientific and economic boundaries. In illustration of this lack of traction, the UK Labour government released a statement in February 2010 confirming that – after hearing medical evidence from the Industrial Injuries Advisory Council (IIAC) and the Chief Medical Officer – it would not be overturning the Law Lords' decision on pleural plaques nor would it be establishing a no-fault compensation scheme.[1] This decision will leave pleural plaque sufferers, such as Alasdair Packard, who was introduced at the beginning of this book, in limbo; knowing they have incurred a disease through their specific form of employment, but having no political process for redress. Rather than recognizing the social and environmental factors that shape people's experience of ARDs, and pleural plaques in particular, the Labour government said it would invest in new informational materials – presenting 'consistent' and 'accurate' information – that aim to, but are in effect highly unlikely to, reassure pleural plaque sufferers. Instead, the Labour government adopted a conservative approach that left powerful economic and political vested interests unchallenged, encouraging insurers to fund increased research into ARDs and improving legal mechanisms for people suffering from mesothelioma.

Asbestos is a microscopic fibre not always visible to the human eye. It has no smell. It comes in different colours and has been embedded in a myriad of different materials. Exposure to asbestos does not create any immediate physical sensation. There are no warning signs prior to exposure and no treatment after exposure. For these reasons, asbestos cannot be managed remotely through technocratic decisions regarding risk. Managing asbestos, protecting against exposure, facilitating environmental rehabilitation and making sure that the most vulnerable people are protected by policy involve new approaches. Building pathways to sustainability, finding ways to live with asbestos-polluted environments and for policy to incorporate an awareness of people's concerns and anxieties, requires a combination of mobilization, political will, activism, serendipity, social networks and taking advantage of 'interconnected spaces'. This also involves developing new skills, talking to and moving 'between' science and society, and understanding how technical systems are influenced – and in turn influence – social processes. Building pathways to sustainability also requires awareness of the complex interactions between the environment, society and technology; of the ways in which policy impacts on diverse people's lives; of the complexity of everyday lives; and of how dominant medical and legal understandings frame asbestos issues. What this book conclusively demonstrates is that, as long as dominant framings prioritize certain economic and political instruments while simultaneously narrowing the possibilities for dealing with asbestos to technical notions of risk, they cannot achieve sustainability and social justice.

Notes

1   The new coalition government seems likely to follow the same policy as its predecessor on this issue.

References

Abelson, P. H. (1990) 'The asbestos removal fiasco', *Science*, vol 247, no 4946, p1017

Abercrombie, N. and Hill, S. (1976) 'Paternalism and patronage', *British Journal of Sociology*, vol 27, no 4, pp413–429

Abratt, R. P., Vorobiof, D. A. and White, N. (2004) 'Asbestos and mesothelioma in South Africa', *Lung Cancer*, vol 45, suppl 1, ppS3–6, www.ncbi.nlm.nih.gov/pubmed/15261425, accessed 28 November 2010

Abu-Lughod, L. (1991) 'Writing against culture', in R. G. Fox (ed) *Recapturing Anthropology: Working in the Present*, School of American Research Press, Sante Fe, NM

Acharya, B. (1989) 'Occupational health and safety in India legislations inadequate', *Indian Journal of Public Administration*, vol 35, pp582–591

Adam, B., Beck, U. and van Loon, J. (2000) *The Risk Society and Beyond: Critical Issues for Social Theory*, Sage, London

Ahn, Y.-S. (2009) 'Environmental fallout from asbestos pollution in Korea: Asbestosis epidemics from neighborhood exposure in Chungnam provinces, Korea', paper presented at the Asian Asbestos Conference, Hong Kong, 26–27 April 2009

Allen, A. (2003) 'Environmental planning and management of the peri-urban interface: Perspectives on an emerging field', *Environment and Urbanization*, vol 15, no 1, pp135–147

Allen, A., Dávila, J. D. and Hofmann, P. (2006) *Governance of Water and Sanitation Services for the Peri-Urban Poor: A Framework for Understanding and Action in Metropolitan Regions*, Development Planning Unit, University College London, London

Anderson, B. (1983) *Imagined Communities: Reflections on the Origin and Spread of Nationalism*, Verso, London

ANROAV (Asian Network for the Rights of Occupational Accident Victims) (2007) *Ban Asbestos Campaign in India*, www.anroav.org/documents/anroav-2007-presentations/BAN%20ASBESTOS%20CAMPAIGN%20IN%20INDIA.doc, accessed 5 September 2007

Ansari, F. A., Ahmad, I., Ashquin, M., Yunus, M. and Rahman, Q. (2007) 'Monitoring and identification of airborne asbestos in unorganized sectors, India', *Chemosphere*, vol 68, pp716–723

Appadurai, A. (1990) 'Technology and the reproduction of values in rural western India', in F. A. Marglin and S. A. Marglin (eds) *Dominating Knowledge: Development, Culture and Resistance*, Claredon Press, Oxford

Appadurai, A. (2000) 'Grassroots globalization and the research imagination', *Public Culture*, vol 12, no 1, pp1–19

Appadurai, A. (2006) 'Foreword', in S. Batliwala and D. L. Brown (eds) *Transnational Civil Society: An Introduction*, Kumarian Press, Bloomfield, CT

Asbestos Project Prieska (n.d.) unpublished preliminary results, Department of Health, Welfare and Environmental Affairs

Audit Commission (2007) *Corporate Assessment: London Borough of Barking and Dagenham*, Audit Commission, London, www.audit-commission.gov. uk/localgov/audit/corpassess/Pages/londonboroughofbarkinganddagenham-corporateassessment.aspx, accessed 28 June 2010

Balshem, M. (1991) 'Cancer, control, and causality: Talking about cancer in a working-class community', *American Ethnologist*, vol 18, no 1, pp152–172

Barking and Dagenham Council (2002) *Improving Health and Wellbeing Through Public Health Partnership*, Annual Report 2002/2003

Bartrip, P. (2001) *The Way from Dusty Death: Turner & Newall and the Regulation of Occupational Health in the British Asbestos Industry, 1890s–1970*, Athlone Press, London

Bartrip, P. (2006) 'Asbestos and health in twentieth century Britain: Motives and outcomes', XIV International Economic History Congress, Helsinki 2006, Session 47, Risks at Work in Europe: Perception, Repair and Prevention (18th–20th Centuries), www.helsinki.fi/iehc2006/papers2/Bartrip.pdf, accessed 13 July 2010

Batliwala, S. and Brown, D. (2006) 'Introduction: Why transnational civil society matters', in S. Batliwala and D. Brown (eds) *Transnational Civil Society: An Introduction*, Kumarian Press, Bloomfield, CT

BBC News (2006) '"Asbestos risk" for India workers', 6 September, http://news.bbc. co.uk/1/hi/world/south_asia/5320386.stm, accessed 14 July 2010

BBC News (2009) 'Asbestos in majority of schools', 27 January, http://news.bbc. co.uk/1/hi/england/7854316.stm, accessed 19 February 2009

Beck, U. (1992) *Risk Society: Towards a New Modernity*, Sage, London

Becker, G. (2000) *The Elusive Embryo: How Women and Men Approach New Reproductive Technologies*, University of California Press, Berkeley, CA

Bernstein, D. and Hoskins, J. A. (2006) 'The health effects of chrysotile: Current perspectives based upon recent data', *Regulatory Toxicology Pharmacology*, vol 45, no 3, pp252–264

Bernstein, D., Rogers, R. and Smith, P. (2003) 'The biopersistence of Canadian chrysotile asbestos following inhalation', *Inhalation Toxicology*, vol 15, no 13, pp1247–1274

Bernstein, D., Rogers, R. and Smith, P. (2005) 'The biopersistence of Canadian chrysotile asbestos following inhalation: Final results through one year after cessation of exposure', *Inhalation Toxicology*, vol 17, no 1, pp1–14

Bignon, J. (1989) 'Introduction: Mineral fibres in the non-occupational environment', in J. Bignon, J. Peto and R. Saracci (eds) *Non-Occupational Exposure to Mineral Fibres*, World Health Organization and International Agency for Research on Cancer, IARC Scientific Publications, no 90

Blandy, F. (2006) 'Asbestos ban "used for economic gain"', *Independent Online*, 7 December, www.iol.co.za/news/south-africa/asbestos-ban-used-for-economic-gain-1.285193 accessed 18 February 2008

Bohme, S. R., Zorabedian, J. and Egilman, D. S. (2005) 'Maximising profit and engendering health: Corporate strategies to avoid litigation and regulation', *International Journal of Occupational and Environmental Health*, vol 11, no 4, pp338–348

Booker, C. (2009) 'Health and Safety Executive asbestos ads were wilfully misleading', *Daily Telegraph*, 3 October, www.telegraph.co.uk/comment/6258079/Health-and-Safety-Executive-asbestos-ads-were-wilfully-misleading.html, accessed 23 June 2010

Botha, J. L., Irwig, L. M. and Strebel, P. M. (1986) 'Excess mortality from stomach cancer, lung cancer, and asbestosis and/or mesothelioma in crocidolite mining districts in South Africa', *American Journal of Epidemiology*, vol 123, no 1, pp30–40

Bourke, J. (2005) *Fear: A Cultural History*, Virago Press, London

Braun, L. and Kisting S. (2006) 'Asbetos-related disease in South Africa: The social production of an invisible epidemic', *American Journal of Public Health*, vol 96, no 8, pp2–12

Braun, L., Greene, A., Manseu, M., Singhal, R., Kisting, S. and Jacobs, N. (2003) 'Scientific controversy and asbestos: Making disease invisible', *International Journal of Occupational and Environmental Health*, vol 9, pp194–205

Bridle, J. and Stone, S. (n.d.) 'Casitile, The new asbestos: Time to clear the air and save £20 billion', www.asbestoswatchdog.co.uk/Science/Casitile,%20The%20New%20Asbestos.pdf, accessed 5 January 2010

Brophy, J. (2006) 'The public health disaster Canada chooses to ignore', in *Chrysotile Asbestos: Hazardous to Humans, Deadly to the Rotterdam Convention*, Building and Woodworkers International and International Ban Asbestos Secretariat, London

Brown, E. R. (1979) *Rockefeller Medicine Men: Medicine and Capitalism in America*, University of California Press, Berkeley/Los Angeles, CA

Brunsson, N. (2003) 'Organized hypocrisy', in B. Czarniawska and G. Sevon (eds) *The Northern Lights*, Copenhagen Business School Press, Copenhagen

Burdett, G. (2006) *Investigation of the Chrysotile Fibres in an Asbestos Cement Sample*, Health and Safety Laboratory, UK, HSL/2007/11, Crown Copyright

Burdett, G. J., Jaffrey, S. A. M. T. and Rood, A. P. (1989) 'Airborne asbestos fibre levels in buildings: A summary of measurements', in J. Bignon, J. Peto and R. Saracci (eds) *Non-Occupational Exposure to Mineral Fibres*, World Health Organization and International Agency for Research on Cancer, IARC Scientific Publications, no 90

Burnham, J. C. (1982) 'American medicine's Golden Age: What happened to it?', *Science*, vol 215, no 4539, pp1474–1479

Cancer Research UK (n.d.) *Mesothelioma Risks and Causes*, www.cancerhelp.org.uk/help/default.asp?page=4395, accessed 22 December 2007

Carnie, T. (2007) 'South African workplaces are like "slow poison"', *Mercury*, 24 July

Castleman, B. (2001) 'Controversies at international organisations over asbestos industry influence', *International Journal of Health Services*, vol 31, no 1, pp193–202

Castleman, B. (2002) *India: Asbestos Update,* 25 April, International Ban Asbestos Secretariat, www. ibasecretariat.org/bc_india_asb_update.php, accessed 28 November 2010

Castleman, B. (2003) 'Commentary: "Controlled use" of asbestos', *International Journal of Occupational and Environmental Health*, vol 9, no 3, pp294–298

Castleman, B. and Lemen, R. (1998) 'Corporate junk science: Corporate influence at international science organisations', *International Journal of Occupational and Environmental Health*, vol 4, pp41–43

Chenoy, K. M. (1985) 'Industrial policy and multinationals in India', *Social Scientist*, vol 13, no 3, pp15–31

Chesson, J., Hatfield, J., Schultz, B., Dutrow, E. and Blake, J. (1990) 'Airborne asbestos in public buildings', *Environmental Research*, vol 51, pp100–107

Chrysotile Institute (2007) *Newsletter: For Safe and Responsible Use of Chrysotile*, vol 6, no 1, pp1–12

Clarke, A. E. and Olesen V. L. (1999) *Revisioning Women, Health and Healing: Feminist Cultural and Technoscience Perspectives*, Routledge, New York, NY

Clayson, H. (2006) 'Mesothelioma: A Patient's Perspective', paper presented at the London School of Hygiene and Tropical Medicine, London, 11 May 2006

Cobb, S. (1976) 'Social support as a moderator of life stress', *Psychosomatic Medicine*, vol 38, pp300–314

Collier, R. (1998) *Masculinities, Crime and Criminology: Men, Heterosexuality and the Criminal(ised) Other*, Sage, London

Commins, B. T. (1991) 'Asbestos in indoor air and its significance: Identification of possible control requirements', *Indoor and Built Environment*, vol 1, p144

Competition Commission (1973) 'Asbestos and certain asbestos products: A report on the supply of asbestos and certain asbestos products series, HC 3, 1972–73', www.competition-commission.org.uk/rep_pub/reports/1970_1975/fulltext/066c11.pdf, accessed 28 July 2010

Connell, R. (1987) *Gender and Power*, Polity Press, Cambridge

Connell, R. and Messerschmidt, J. W. (2005) 'Hegemonic masculinity: Rethinking the concept', *Gender and Society*, vol 19, no 6, pp829–859

Coombs, A. (2002) 'Three minerals, three epidemics', paper presented in Australia, November 2002, www.johnpickering.co.uk/index.php?option=com_content&view=article&id=51:27th-november-2002-three-minerals-three-epidemics&catid=1:news&Itemid=6, accessed 9 August 2004

Cornwall, A. (2004) 'New democratic spaces: The politics and dynamics of institutionalised participation', *IDS Bulletin*, vol 35, no 2, pp1–10

Craighead, J. E. (2008) 'Diseases associated with asbestos industrial products and environmental exposure', in J. E. Craighead and A. R. Gibbs (eds) *Asbestos and Its Diseases*, Oxford University Press, Oxford

Cullinan, K. (2002) 'Man sues asbestos giant over rare cancer', *Health-e News Service*, www.health-e.org.za/news/article.php?uid=20020117, accessed 2 March 2010

Daubs, K. (2008) 'Canada's asbestos time bomb. Part III: India's hidden epidemic', *Ottawa Citizen*, 24 October, www.canada.com/ottawacitizen/features/asbestos/story.html?id=e7fb6e00-c03f-4a5c-b92e-6205e68bf816, accessed 14 March 2009

Davies, J. C., Kielkowski, D., Phillips, J. I., Govuzela, M., Solomon, A., Makofane, M. R., Sekgobela, M. L. and Garton, E. (2004) 'Asbestos in the sputum, crackles in the lungs and radiological changes in women exposed to asbestos', *International Journal of Occupational and Environmental Health*, vol 10, pp220–225

Department of Environmental Affairs and Tourism (2008a) 'Asbestos Regulations, 2008', *Waste Information Today*, no 2, June/July, pp1–3, http://wis.octoplus.co.za/documents/338.pdf, accessed 28 June 2010

Department of Environmental Affairs and Tourism (2008b) Statement by the office of Marthinus van Schalkwyk, Minister of Environmental Affairs and Tourism at the launch of the Asbestos Regulations, Pretoria, 27 March 2008, www.info.gov.za/speeches/2008/08032813451004.htm, accessed 28 December 2009

Department of Health, Welfare and Environmental Affairs (n.d.) 'Preliminary results: Asbestos study: Prieska', unpublished document

Department of Minerals and Energy (2008) *Annual Report, 2007/2008, Part 2*, www.info.gov.za/view/DownloadFileAction?id=93531, accessed 29 October 2010

De Saulles, D. (2006a) 'The very seamark of my utmost sail: Grieves and the continuing asbestos wars', *Journal of Personal Injury Law*, vol 2, pp119–161

De Saulles, D. (2006b) 'The asbestos wars: The American experience', *Journal of Personal Injury Law*, vol 3, pp215–250

De Saulles, D. (2006c) 'Nought plus nought plus nought equals nought: Rhetoric and the asbestos wars', *Journal of Personal Injury Law*, vol 4, pp301–336

De Sousa Santos, B. and Rodriguez-Garavito, C. A. (2005) *Law and Globalisation from Below: Towards a Cosmopolitan Legality*, Cambridge University Press, Cambridge

De Vuyst, P. and Genevois, P. A. (2002) 'Asbestosis', in D. Hendrick, W. Beckett, P. S. Burge and A. Churg (eds) *Occupational Disorders of the Lung: Recognition, Management and Prevention*, Saunders Ltd, London

DGFASLI (Directorate General Factory Advice Service and Labour Institutes) (2004) *Report on the National Study on Health Status of Workers in the Asbestos Industry*, DGFASLI , Government of India, Ministry of Labour and Employment

DGFASLI (2005) *National Study on Status of Work Environment in Asbestos Products Manufacturing Industry: Evaluation of Airborne Asbestos Fibres*, Industrial Hygiene Division, Central Labour Institute, DGFASLI, Government of India, Ministry of Labour and Employment

DGFASLI (2006) 'Hazards of Asbestos and Silica in Construction Industry', report on a national seminar, http://dgfasli.nic.in/seminar/proceedings.htm, accessed 28 June 2010

'Distrikstrekordboek' (1978) Unpublished scrapbook, initiated 1978, no title, compiled by successive Griquatown magistrates

Doll, R. (1955) 'Mortality from lung cancer in asbestos workers', *British Medical Journal*, vol 12, pp81–86

Douglas, M. and Wildavsky, A. B. (1982) *Risk and Culture: An Essay on the Selection of Technical and Environmental Dangers*, University of California Press, Berkeley, CA

Dressler, W. W. (1980) 'Coping dispositions, social supports, and health status', *Ethnos*, vol 8, no 2, pp146–171

Dunnigan, J. (1993) 'Chrysotile asbestos revisited', *British Journal of Industrial Medicine*, vol 50, pp862–863

Dupont, V. (2005) 'Peri-urban dynamics: Population, habitat and environment on the peripheries of large Indian metropolises: An introduction', in V. Dupont (ed) *Peri-Urban Dynamics: Population, Habitat and Environment on the Peripheries of Large Indian Metropolises: A Review of Concepts and General Issues,* CHS Occasional Paper No 14, French Research Institutes India

Durodié, B. (n.d.) *Risk Case Studies: The Concept of Risk,* Nuffield Trust Global Programme on Health, Foreign Policy and Security, Nuffield Health and Social Services Fund, UK Global Health Fund

Dutta, M. (2008a) *A Fox in the Hen House: Made-to-order Science and India's Asbestos Policy,* Corporate Accountability Desk – The Other Media, www.bwint.org/pdfs/Asbestos%20Dossier%20final%20Feb%202008.pdf, accessed 2 May 2009

Dutta, M. (2008b) 'The Indian government's complicity in the asbestos scandal', in D. Allen and L. Kazan-Allen (eds) *India's Asbestos Time Bomb*, International Ban Asbestos Secretariat, London

Dutta, M. (2009) 'India – country report', paper presented at the Asian Asbestos Conference, Hong Kong, 26–27 April 2009

Edelman, M. (2001) 'Social movements: Changing paradigms and forms of politics', *Annual Review of Anthropology*, vol 30, pp285–317

Edwards, J. (2000) *Born and Bred: Idioms of Kinship and New Reproductive Technologies in England*, Oxford University Press, Oxford

Edwards, J., Harvey, P. and Wade, P. (2007) 'Introduction: Epistemologies in practice', in J. Edwards, P. Harvey and P. Wade (eds) *Anthropology and Science: Epistemologies in Practice*, Berg Publishers, Oxford

Egilman, D. S. (2003) 'Exposing the "myth" of ABC, "Anything But Chrysotile": A critique of the Canadian asbestos mining industry and McGill University chrysotile studies', *American Journal of Industrial Medicine*, vol 44, pp540–557

Egilman, D. S. and Billings, M. A. (2005) 'Abuse of epidemiology: Automobile manufacturers manufacture a defence to asbestos liability', *International Journal of Occupational and Environmental Health*, vol 11, no 4, pp360–371

Egilman, D. S. and Bohme, S. R. (2005) 'Over a barrel: Corporate corruption of science and its effects on workers and the environment', *International Journal of Occupational and Environmental Health*, vol 11, no 4, pp331–337

Erikson, K. (1990) 'Toxic reckoning: Business faces a new kind of fear', *Harvard Business Review*, vol 68, no 1, pp118–126

Escobar, A. (1999) 'Analyses and interventions: Anthropological engagements with environmentalism', *Current Anthropology*, vol 40, no 3, pp291–293

Escobar, A. and Alvarez, S. E. (eds) (1992) *The Making of Social Movements in Latin America: Identity, Strategy and Democracy,* Westview, Boulder, CO

Evans-Pritchard, E. E. (1937) *Witchcraft, Oracles and Magic Among the Azande*, Oxford University Press, Oxford

Eversheds Briefing (2006) 'Landmark court decision chips away at asbestos compensation claims', *Eversheds Briefing*, 23 February, www.eversheds.com/Documents/

InsuranceBriefingAsbestos23Feb06(8705).pdf#search=%22landmark%20court %20decision%20eversheds%22, accessed 28 June 2010

Fairhead, J. and Leach, M. (2003) *Science, Society and Power: Environmental Knowledge and Policy in West Africa and the Caribbean*, Cambridge University Press, Cambridge

Falk, R. (1994) 'The making of global citizenship', in B. van Steenbergen (ed) *The Condition of Citizenship*, Sage, London

Faulkner, W. (2000) 'The power and the pleasure? A research agenda for "making gender stick" to engineers', *Science, Technology and Human Values*, vol 25, no 1, pp87–119

Felix, M. A. (1991) 'Risking their lives in ignorance: The story of an asbestos-polluted community', in J. Cock and E. Koch (eds) *Going Green: People, Politics and the Environment in South Africa*, Oxford University Press, Oxford

Felix, M. A., Leger, J.-P. and Ehrlich, R. I. (1994) 'Three minerals, three epidemics: Asbestos mining and disease in South Africa', in M. A. Melman and A. Upton (eds) *The Identification and Control Of Environmental and Occupational Diseases; Asbestos and Cancers*, vol XXIII, Princeton Scientific Publishing Co, Princeton, NJ

Felstiner, W. L. F. and Dingwall, R. (1988) *Asbestos Litigation in the United Kingdom: An Interim Report*, Centre for Socio-Legal Studies, Oxford

Field Fisher Waterhouse (n.d.) *Mesothelioma: A Patient's Guide to Compensation*, Field Fisher Waterhouse, London

Field Fisher Waterhouse (2006) *Asbestos Disease Claims: A Client's Guide*, Field Fisher Waterhouse, London

Fischer, F. (2005) 'Are scientists irrational? Risk assessment in practical reason', in M. Leach, I. Scoones and B. Wynne (eds) *Science and Citizens: Globalization and the Challenge of Engagement*, Zed Books, London

Flynn, T. G. (1999) 'The two cultures: Forty years later', in A. M. Herzberg and I. Krupka (eds) *Statistics, Science and Public Policy IV: The Two Cultures?*, proceedings of the conference on Statistics, Science and Public Policy, Hailsham, UK, 21–24 April 1999

Foucault, M. (1991) 'On governmentality', in G. Burchell, C. Gordon and P. Miller (eds) *The Foucault Effect*, Harvester Wheatsheaf, London

Fox, J. (2005) 'Unpacking "transnational citizenship"', *Annual Review of Political Science*, vol 8, pp171–201

Franklin, S. (1995) 'Science as culture, cultures of science', *Annual Review of Anthropology*, vol 24, pp163–184

Franklin, S. and Lock, M. (2003) *Remaking Life and Death: Toward an Anthropology of the Biosciences*, SAR Press, Santa Fe, NM

Franklin, S. and McKinnon, S. (eds) (2001) *Relative Values: Reconfiguring Kinship*, Duke University Press, Durham, NC

Franklin, S. and Ragoné, H. (1998) *Reproducing Reproduction: Kinship, Power and Technological Innovation*, University of Pennsylvania Press, Philadelphia, PA

Frazer, J. (1911) *The Golden Bough: A Study in Comparative Religion*, Macmillan, London

Furedi, F. (1997/2002) *Culture of Fear: Risk-taking and the Morality of Low Expectation*, Continuum, London

Gaitonde, R. and Dutta, M. (2008) 'The struggle by Mumbai workers for compensation', in D. Allen and L. Kazan-Allen (eds) *India's Asbestos Time Bomb*, International Ban Asbestos Secretariat, London

Ganguly, P. K. (2007) 'Dumping of toxic waste in Asia: Ship-breaking in India', in L. Kazan-Allen, *Killing the Future: Asbestos Use in Asia*, International Ban Asbestos Secretariat, London

Gardener, M. J. and Powell, A. (1986) 'Mortality of asbestos cement workers using almost exclusively chrysotile fibre', *Journal of the Society of Occupational Medicine*, vol 36, no 4, pp124–126

Gaventa, J. (2002) 'Introduction: Exploring citizenship, participation and accountability', *IDS Bulletin*, vol 33, no 2, pp1–11

Gee, D. and Greenberg M. (2002) 'Asbestos: From "magic" to malevolent mineral', in P. Harremo, D. Gee, M. MacGavin, A. Stirling, J. Keys, B. Wynne and S. Guedez Vaz (eds) *Late Lessons from Early Warnings: The Precautionary Principle 1896–2000*, European Environment Agency, Copenhagen

Gibson, B. (2007a) 'The spirit of the summit: A brief review of Everite's new technology campaign', unpublished presentation, 11 June 2007

Gibson, B. (2007b) 'Brian Gibson issue management', report on a meeting of the Chrysotile Joint Technical Task Force Team, Cape Town, 20–21 June 2007

Gilmore, D. (1990) 'Men and women in southern Spain: "Domestic power revisited"', *American Anthropologist*, vol 92, no 4, pp953–970

Good, B. J. (1994) *Medicine, Rationality, and Experience: An Anthropological Perspective*, Cambridge University Press, Cambridge

Gordon, L. (1990) 'The new feminist scholarship on the welfare state', in L. Gordon (ed) *Women, the State and Welfare*, University of Wisconsin Press, Madison, WI

Government Communication and Information System (GCIS) (2009) *South Africa Yearbook 2008/09*, GCIS, Pretoria, www.gcis.gov.za/resource_centre/sa_info/yearbook/2008-09.htm, accessed 29 October 2010

Gqubule, T. (2006) 'White asbestos: Zimbabwe's great white hope', *Financial Mail*, 3 November, p85

Gravelsons, B., Hawes, W., Jakubowski, S., Kent, A., Lowe, J., Macnair, A., Michaels, D., Morton, A., Sanders, D., Towell, P., Whiting, A., Widdows, J. and Williams, A. (2004) *UK Asbestos: The Definitive Guide*, www.actuaries.org.uk/__data/assets/pdf_file/0004/34969/Lowe.pdf, accessed 28 June 2010

Grieves and others v F. T. Everard & Sons and British Uralite Plc and others [2005] EWHC 88 (QB)

Grieves and others v F. T. Everard & Sons and others [2006] EWCA Civ 27 (transcript)

Groenewald, Y. (2005) 'Deadly dust ban', *Mail & Guardian*, 31 August

Guidotti, T. L. (1988) 'Quantitative risk assessment of exposure to airborne asbestos in an office building', *Canadian Journal of Public Health*, vol 79, pp249–254

Gupta, A. and Ferguson, J. (1992) 'Beyond "culture": Space, identity and the politics of difference', *Cultural Anthropology*, vol 7, no 1, pp6–23

Gutmann, M. C. (1997) 'Trafficking in men: The anthropology of masculinity', *Annual Review of Anthropology*, vol 26, pp385–409

Hacker, S. (1989) *Pleasure, Power and Technology: Some Tales of Gender, Engineering and Cooperative Workplace*, Unwin Hyman, Boston, MA

Halford, S. J. (1949) *The Griquas of Griqualand: A Historical Narrative of the Griqua People: Their Rise, Progress, and Decline*, Juta and Company, Cape Town

Hall, A. L. (1930) *Asbestos in the Union of South Africa. Geological Survey Memoir No 12*, Department of Mines and Industries, Pretoria

Hall, S. (1989) 'Ethnicity: Identity and difference', *Radical America*, vol 23, no 4, pp9–20

Hammond, C. E., Selikoff, I. J. and Seidman H. (1979) 'Asbestos exposure, cigarette smoking and death rates', *Annals of the New York Academy of Sciences*, vol 330, no 1, pp473–790

Handwerker, L. (2003) 'New genetic technologies and their impact on women: A feminist perspective', *Gender and Development*, vol 11, no 1, pp114–124

Harding, S. (1996) 'Science is "good to think with"', *Social Text*, no 46/47, pp15–26

Harding, S. (1998) *Is Science Multicultural? Postcolonialisms, Feminisms, and Epistemologies*, Indiana University Press, Bloomington, IN

Hearn, J. I. and Morgan, D. H. (1990) 'Men, masculinities and social theory', in J. I. Hearn and D. H. Morgan (eds) *Men, Masculinities and Social Theory: Critical Studies on Men and Masculinities*, Unwin Hyman, London

Heater, D. (2002) *World Citizenship: Cosmopolitan Thinking and its Opponents*, Continuum, London

Held, D. and McGrew, A. (2002) 'Introduction', in D. Held and A. McGrew (eds) *Governing Globalization: Power, Authority and Global Governance*, Polity Press, Cambridge

Heller, D. S., Gordon, R. E., Westhoff, C. and Gerber, S. (1996) 'Asbestos exposure and ovarian fiber burden', *American Journal of Industrial Medicine*, vol 29, no 5, pp435–439

Helman, C. (1984) *Culture, Health and Illness*, Wright PSG, London

Henwood, K., Parkhill, K. and Pidgeon, N. (2006) 'Science, technology and risk perception: From gender differences to the effects made by gender', paper presented at GEM-SET conference, University of Newcastle, 23 November 2006

Hessel, P. A. and Sluis-Cremer, G. K. (1989) 'X-ray findings, lung function, and respiratory symptoms in black South African vermiculite workers', *American Journal of Industrial Medicine*, vol 15, no 1, pp21–29

Hirway, I. and Mahadevia, D. (2004) *Gujarat Human Development Report 2004*, Mahatma Gandhi Labour Institute, Ahmedabad

Hopley, M. J. and Richards, G. A. (1999) 'Asbestos-related disease: A community-based and a capture-recapture analysis', Respiratory Units, Chris Hani Baragwanath Hospital and the Johannesburg Hospital, University of the Witwatersrand, unpublished study

HSE (Health and Safety Executive) (2006) 'The management of asbestos in non-domestic premises', Regulation 4 of the Control of Asbestos Regulations 2006, Crown Copyright, UK

HSRC (Human Sciences Research Council) (2004) 'Fact sheet: Poverty in South Africa', 26 July, www.sarpn.org/documents/d0000990, accessed 29 June 2010

Hughes, J. M., Weill, H. and Hammad, Y. Y. (1987) 'Mortality of workers employed in two asbestos cement manufacturing plants', *British Journal of Industrial Medicine*, vol 44, no 3, pp161–174

IHRC (Industrial Health Research Group) (1996) 'Report on the audit of the asbestos-related disease surveillance programme at GEFCO crocidolite asbestos mine conducted on 20 and 21 May 1996', unpublished report

IIAC (Industrial Injuries Advisory Council) (2009) *Pleural Plaques*, Position Paper 23, www.iiac.org.uk/pdf/pos_papers/pp23.pdf, accessed 29 June 2010

Iida, K. (2009) 'Dust prevention for carpenters: Grassroots activity on a construction site in Tokyo', paper presented at the Asian Asbestos Conference, Hong Kong, 26–27 April 2009

ILO (International Labour Organization) (2006) 'ILO adopts new measures on occupational safety and health, the employment relationship, asbestos', press release ILO/06/34, 15 June, www.ilo.org/global/About_the_ILO/Media_and_public_information/Press_releases/lang--en/WCMS_070506/index.htm, accessed 6 January 2010

Jacobs, N. J. (2003) *Environment, Power, and Injustice: A South African History*, Cambridge University Press, Cambridge

Jasanoff, S. (1995) *Science at the Bar*, Harvard University Press, Cambridge, MA

Jasanoff, S. (1996) 'Beyond epistemology: Relativism and engagement in the politics of science', *Social Studies of Science*, vol 26, no 2, pp393–418

Jasanoff, S. (2003) 'Breaking the waves in science studies: Comment on H. M. Collins and Robert Evans, "The Third Wave of Science Studies"', *Social Studies of Science*, vol 33, no 3, pp389–400

Jasanoff, S. (2004) 'The idiom of co-production', in S. Jasanoff (ed) *States of Knowledge: The Co-production of Science and Social Order*, Routledge, London

Jasanoff, S. (2005) *Designs on Nature*, Princeton University Press, Princeton, NJ

Jasanoff, S. (2009) 'Governing Innovation', *Seminar 597 Knowledge in Question*, May 2009, pp16–25

Joshi, T. K., Bhuva, U. B and Katoch, P. (2006) 'Asbestos ban in India: Challenges ahead', *Annals New York Academy of Sciences*, vol 1076, pp292–308

Joubert, P. (2007) 'A lethal find', *Mail & Guardian*, 30 November–6 December

Kabeer, N. (2002) *Citizenship and the Boundaries of the Acknowledged Community: Identity, Affiliation and Exclusion*, IDS Working Paper 171, Institute of Development Studies, Brighton

Kamat, S. R. (2008) 'Asbestos-related disease in India', in D. Allen and L. Kazan-Allen (eds) *India's Asbestos Time Bomb*, International Ban Asbestos Secretariat, London

Kasperson, R. E. and Kasperson, J. X. (2005) 'Hidden hazards', in R. E. Kasperson and J. X. Kasperson (eds) *The Social Contours of Risk: Volume I: Publics, Risk Communication and the Social Amplification of Risk*, Earthscan, London

Kasperson, R. E. and Pijawka, K. D. (2005) 'Societal response to hazards and major hazard events: Comparing natural and technological disasters', in R. E. Kasperson and J. X. Kasperson (eds) *The Social Contours of Risk: Volume II: Risk Analysis, Corporations and the Globalizaation*, Earthscan, London

Kazan-Allen, L. (n.d.) 'South Africa: The asbestos legacy', International Ban Asbestos Secretariat, www.ibasecretariat.org/lka_sa_leg.php, accessed 29 June 2010

Kazan-Allen, L. (1999) 'Asbestos finally banned in the United Kingdom', *British Asbestos Newsletter*, issue 36, autumn 1999, www.ibasecretariat.org/lka_uk_ban.php, accessed 30 June 2010

Kazan-Allen, L. (2002) 'Recent developments affecting European asbestos claims', International Ban Asbestos Secretariat, www.ibasecretariat.org/lka_bahia_2_txt.php, accessed 2 March 2010

Kazan-Allen, L. (2003a) 'Canadian asbestos: A global concern', International Ban Asbestos Secretariat, www.ibasecretariat.org/lka_ottawa_conf_rep_03_plus.php, accessed 29 June 2010

Kazan-Allen, L. (2003b) 'The asbestos war', *International Journal of Occupational and Environmental Health*, vol 9, no 3, pp173–193

Kazan-Allen, L. (2005) 'Asbestos and mesothelioma: Worldwide trends', *Lung Cancer*, vol 49, ppS3–S8

Kazan-Allen, L. (2006) 'Chronological record of the contributions of national delegations and others', in *Chrysotile Asbestos: Hazardous to Humans, Deadly to the Rotterdam Convention*, Building and Woodworkers International and International Ban Asbestos Secretariat, London

Kazan-Allen, L. (2007) 'Two steps forward, five steps back', International Ban Asbestos Secretariat, www.ibasecretariat.org/lka_2_steps_forward_5_back.php, accessed 27 January 2010

Kazan-Allen, L. (2008) *British Asbestos Newsletter*, issue 71, pp1–6, summer 2008

Kazan-Allen, L. (2009a) *British Asbestos Newsletter*, issue 75, pp1–6, summer 2009

Kazan-Allen, L. (2009b) *Asian Asbestos Conference 2009*, International Ban Asbestos Secretariat in conjunction with Asia Monitor Resource Center, Asian Network for the Rights of Occupational Accident Victims, Building and Woodworkers International, Corporate Accountability Desk – The Other Media, India, Association for the Rights of Industrial Accident Victims, Hong Kong Confederation of Trade Unions, Japan Association of Mesothelioma and Asbestos-Related Disease Victims and their Families, Ban Asbestos Network Japan, Ban Asbestos Network Korea

Keck, M. and Sikkink, K. (1998) *Activist Beyond Borders: Advocacy Networks in International Politics*, Cornell University Press, London

Keegan, J. (2009) 'Action plea on asbestos', *Manchester Evening News*, 13 February, www.manchestereveningnews.co.uk/news/s/1096435_action_plea_on_asbestos, accessed 19 February 2009

Keohane, R. and Nye Jr, J. S. (2000) 'Introduction', in J. S. Nye Jr and J. Donahue (eds) *Governing in a Globalizing World: Visions of Governance for the 21st Century*, Brookings Institution Press, Washington, DC

Kgalamono, S. M., Rees, D., Kielkowski, D and Solomon, A. (2005) 'Asbestos in the non-mining industry on the Witwatersrand, South Africa', *South African Medical Journal*, vol 95, pp47–51

Kielkowski, D., Nelson, G. and Rees, D. (2000) 'Risk of mesothelioma from exposure to crocidolite asbestos: A 1995 update of a South African mortality study', *Occupational and Environmental Medicine*, vol 57, pp563–567

Kisting, S. (2000) 'Asbestos contaminated land in South Africa: The challenges and possibilities', plenary session, Global Asbestos Congress, Osasco, Brazil, 17–20 September 2000

Kleinman, A. (1974) 'Cognitive structures of traditional medical systems', *Ethnomedicine*, vol 3, pp27–49

Kleinman, A. (1977) 'Depression, somatization and the new cross-cultural psychiatry', *Social Science and Medicine*, vol 11, no 1, pp3–10

Kleinman, A. (1981) *Patients and Healers in the Context of Culture: An Exploration of the Borderland between Anthropology, Medicine and Psychiatry*, University of California Press, Berkeley, CA

Kleinman, A. (1995) *Writing at the Margin: Discourse between Anthropology and Medicine*, University of California Press, Berkeley, CA

Kohlstedt, S. G. and Longino, H. (1997) 'The women, gender and science question: What do research on women in science and research on gender and science have to do with each other?', *Osiris,* 2nd series, vol 12, pp3–15

Komarovsky, M. (1967) *Blue Collar Marriage*, Vintage, New York, NY

Kothari, U. (2005) 'Eurocentrism and postcolonialism', in T. Forsythe (ed) *Encyclopedia of International Development*, Routledge, London

Krishna, G. (2006a) 'Asbestos: Kill the people, protect the industry', *Merinews*, 17 October, www.merinews.com/catFull.jsp?articleID=123573, accessed 30 June 2010

Krishna, G. (2006b) 'India's position on chrysotile asbestos dictated by vested interests', in *Chrysotile Asbestos: Hazardous to Humans, Deadly to the Rotterdam Convention*, Building and Woodworkers International and International Ban Asbestos Secretariat, London

Krishna, G. (2007a) 'India's position on chrysotile asbestos dictated by vested interests!', Mesothelioma Information Resource Group, 20 February, www.mirg.org/2007/02/indias-position-on-chrysotile-asbestos-dictated-by-vested-interests, accessed 3 December 2007

Krishna, G. (2007b) 'Human rights and environmental groups condemn India for callous disregard for its poorest labourers and the law after shipbreaking ruling', http://banasbestosindia.blogspot.com/2007/09/human-rights-environmental-groups.html, accessed 12 July 2010

Laclau, E. and Mouffe, C. (1985) *Hegemony and Socialist Strategy: Towards a Radical Democratic Politics*, Verso, London

LaDou, J. (2004) 'The asbestos cancer epidemic', *Environmental Health Perspectives*, vol 112, no 3, pp285–290

Lancet (2008) 'Editorial: Asbestos-related disease – a preventable burden', 6 December, vol 372, p1927

Landrigan, P. J. and Soffritti, M. (2004) 'Dr T. K. Joshi and asbestos in India: A message from the College Ramazzini', *American Journal of Industrial Medicine*, vol 45, pp125–128

Landrigan, P. J. and Soffritti, M. (2005) 'Collegium Ramazzini call for an international ban on asbestos', *American Journal of Industrial Medicine*, vol 47, pp471–474

Landrigan, P. J., Nicholson, W. J., Suzuki, Y. and LaDou, J. (1999) 'The hazards of chrysotile asbestos: A critical review', *Journal of Industrial Health*, vol 37, no 3, pp271–280

Leach. M., Scoones, I. and Stirling, A. (2007) *Pathways to Sustainability: An Overview of the STEPS Centre Approach*, STEPS Approach Paper, STEPS Centre, Brighton

Leach, M., Scoones, I. and Stirling, A. (2010) *Dynamic Sustainabilities: Technology, Environment, Social Justice*, Pathways to Sustainability Series, Earthscan, London

Leach, M., Scoones, I. and Thompson, L. (2002) 'Citizenship, science and risk: Conceptualising relationships across issues and settings', *IDS Bulletin*, vol 33, no 2, pp40–48

Leach, M., Scoones, I. and Wynne, B. (2005) *Science and Citizens: Globalization and the Challenge of Engagement*, Zed Books, London

Lees, M. (2005, amended 2006) 'Asbestos exposure in schools', www.schoolasbestosaction. co.uk/asbestos%20exposure%20in%20schools.pdf, accessed 14 July 2010

Le Guen, J. M., and Burdett, G. (1981) 'Asbestos concentrations in public buildings: A preliminary report', *Annals of Occupational Hygiene*, vol 24, no 2, pp185–189

Leigh, J. and T. Driscoll (2003) 'Malignant mesothelioma in Australia 1945–2002', *International Journal of Occupational and Environmental Health*, vol 9, no 3, pp206–217

Lemen, R. A. (2004) 'Chrysotile asbestos as a cause of mesothelioma: An application of the Hill causation model', *International Journal of Occupational and Environmental Health*, vol 10, no 2, pp233–239

Lemen, R. A. (2008) 'Smoke and mirrors: Chrysotile asbestos is good for you – illusion and confusion but not a fact', in D. Allen and L. Kazan-Allen (eds) *India's Asbestos Time Bomb*, International Ban Asbestos Secretariat, London

Leprince, A. et al (2007) 'Special commission on prevention: Asbestos: Protecting the future and coping with the past', World Social Security Forum, 29th International Social Security Association General Assembly, http://hesa.etui-rehs.org/uk/dossiers/files/leprince-asbestos.pdf, accessed 9 January 2009

Levi-Strauss, C. (1963) *Totemism*, Beacon Press, Boston, MA

Liotard, K. (2006) 'Preface', in *Chrysotile Asbestos: Hazardous to Humans, Deadly to the Rotterdam Convention*, Building and Woodworkers International and International Ban Asbestos Secretariat, London

Lipschutz, R. (2004) 'Global civil society and global governmentality: Or, the search for politics and the state amidst the capillaries of social power', in M. Barnett and R. Duvall (eds) *Power in Global Governance*, Cambridge University Press, Cambridge

Lohan, M. (2000) 'Constructive tensions in feminist technology studies', *Social Studies of Science*, vol 30, pp895–916

London Hazards Centre (1995) *Asbestos Hazards Handbook*, London Hazards Centre, London

Long, N. (1996) 'Globalization and localization: New challenges to rural research', in H. L. Moore (ed) *The Future of Anthropological Knowledge*, Routledge, London

Long Martello, M. L. and Jasanoff, S. (2004) *Earthly Politics: Local and Global in Environmental Governance*, MIT Press, Cambridge, MA

Lublin, N. (1998) *Pandora's Box: Feminism Confronts Reproductive Technology*, Rowman & Littlefield, New York, NY

Lupton, D. (1999) *Risk*, Routledge, New York, NY

Makoshori, S. (2008) 'Zimbabwe: Asbestos mines face staff exodus', *Financial Gazette*, 12 June, http://allafrica.com/stories/200806130789.html, accessed 1 December 2009

Malavadkar, P., Sakpal, S., Kakade, A. and Murlidhar, V. (2009) 'Asbestosis among ex-asbestos workers of a Mumbai asbestos factory: A prevalence study', paper presented at the Asian Asbestos Conference, Hong Kong, 26–27 April 2009

Malinowski, B. (1944) *The Scientific Theory of Culture*, University of North Carolina Press, NC

Manavar, R. and Patel, M. (2004) *Struggle for Justice: Case study of Asbestos Victims in Gujarat*, India Global Asbestos Congress, 19–21 November 2004, Waseda University, Tokyo, http://worldasbestosreport.org/conferences/gac/gac2004/ws_F_05_e.pdf, accessed 30 June 2010

Marcus, G. E. (1999) 'What is at stake – and is not – in the idea and practice of multi-sited ethnography', *Canberra Anthropology*, vol 22, no 2, pp6–14

Marcus, G. E. and Fischer, M. M. J. (1986) *Anthropology as Cultural Critique: An Experimental Moment in the Human Science*, University of Chicago Press, Chicago, IL

Matambanadzo, P. (2007a) 'SA extends deadline on ban of asbestos use', *Herald*, 25 June

Matambanadzo, P. (2007b) 'Zim gets nod to produce asbestos', *Herald*, 25 June

Maule, M. M., Magnani, C., Dalmasso, P., Mirabelli, D., Merletti, F. and Biggeri, A. (2007) 'Modelling mesothelioma risk associated with environmental asbestos exposure', *Environmental Health Perspectives*, vol 115, no 7, pp1066–1071

McClintick, D. (2000) 'The decline and fall of Lloyd's of London', *Time*, vol 155, no 7, pp36–58

McCormick, S. (2009) *Mobilizing Science: Movements, Participation, and the Remaking of Knowledge*, Temple University Press, Philadelphia, PA

McCulloch, J. (2002) *Asbestos Blues: Labour, Capital, Physicians and the State in South Africa*, James Currey, Oxford

McCulloch, J. (2003) 'Asbestos mining in Southern Africa, 1893–2002', *International Journal of Occupational and Environmental Health*, vol 9, pp230–235

McCulloch, J. (2005) 'Asbestos, lies and the state: Occupational disease and South African science', *African Studies*, vol 64, no 2, pp201–216

McCulloch, J. and Tweedale, G. (2008) *Defending the Indefensible: The Global Asbestos Industry and its Fight for Survival*, Oxford University Press, Oxford

McInnes, C. (2005) *Health, Security and the Risk Society*, Nuffield Trust Global Programme on Health, Foreign Policy and Security, Nuffield Health and Social Services Fund, UK Global Health Fund

Meeran R. (2003) 'Cape plc: South African mineworkers' quest for justice', *International Journal of Occupational and Environmental Health*, vol 9, no 3, pp218–229

Mellström, U. (2002) 'Patriarchal machines and masculine embodiment', *Science, Technology and Human Values*, vol 27, no 4, pp460–478

Menon, R. (2006) 'Psychological effects of an asbestos related disease', presentation at the Mesothelioma Conference, London School of Hygiene and Tropical Medicine, London, 11 May 2006

Michael, M. (2002) 'Comprehension, apprehension, prehension: Heterogeneity and the public understanding of science', *Science, Technology and Human Values*, vol 27, no 3, pp357–378

Ministry of Justice (2008) 'Pleural plaques', consultation paper CP14/08, Ministry of Justice, UK, www.justice.gov.uk/docs/cp1408.pdf, accessed 14 July 2010

Misra, S. (2003) 'Top medical experts want "carcinogenic" asbestos phased out', *Tribune*, 25 April, www.tribuneindia.com/2003/20030425/ncr3.htm, accessed 7 January 2006

Mohanty, R. and Tandon, R. (2005) *Engaging with the State: Implications for Citizenship and Participation*, Citizenship DRC Synthesis Paper

Mohite, R. G. (2008) 'Preface', in D. Allen and L. Kazan-Allen (eds) *India's Asbestos Time Bomb*, International Ban Asbestos Secretariat, London

Mollona, M. (2005) 'Factory, family and neighbourhood: The political economy of informal labour in Sheffield', *Journal of the Royal Anthropological Institute*, vol 11, pp527–548

Monbiot, G. (2008) 'The patron saint of charlatans is again spreading dangerous information', *Guardian*, 23 September, www.guardian.co.uk/commentisfree/2008/sep/23/controversiesinscience.health, accessed 5 January 2010

Moore, H. (1993) *A Passion for Difference: Essays in Anthropology and Gender*, Polity Press, Cambridge

Moore, H. (1996) 'The changing nature of anthropological knowledge: An introduction', in H. Moore (ed) *The Future of Anthropological Knowledge*, Routledge, London

Moore, S. and Lenaghan, J. (1995) 'Invisible Death', *New Statesman and Society*, 10 February

Morris, R. (2004) 'Asbestosis victims continue their fight while multinationals duck and dive', *Business Report*, 8 January, www.business-humanrights.org/Links/Repository/919449/link_page_view, accessed 1 July 2010

Mossman, B. T. and Gee, B. L. (1989) 'Asbestos-related diseases', *New England Journal of Medicine*, vol 320, pp1721–1730

Murlidhar, V. and Kanhere, V. (2005) 'Asbestosis in an asbestos composite mill at Mumbai: A prevalence study', *Environmental Health: A Global Access Science*, vol 4, p24

Murphree, D. W., Wright, S. and Ebaugh, H. R. (1996) 'Toxic waste siting and community resistance: How cooptation of local citizen opposition failed', *Sociological Perspectives*, vol 39, no 4, pp447–463

Myers, J. (1981) 'The social context of occupational disease: Asbestos and South Africa', *International Journal of Health Services*, vol 11, no 2, pp227–245

Nath, M. (2000) 'Industrial disasters: Working towards oblivion', *Other Disasters: India Disasters Report*, www.indiadisasters.org/idrpdf/other%20Disasters/otherdisasters%20Industrial,pdf, accessed 3 December 2007

NCDRLD (National Campaign on Dust-Related Lung Diseases) (1992) *An Untold Story: The Ongoing Struggle of Textile Workers in Ahmedabad*, Ashish Printers and Publication, New Delhi

Newell P. (2001) 'Managing multinationals: The governance of investment for the environment', *Journal of International Development*, vol 13, pp907–919

Nicholson, W. J. (2001) 'The carcinogenicity of chrysotile asbestos: A review', *Industrial Health*, vol 39, no 2, pp57–64

NIOH (National Institute of Occupational Health) (1997) *A Report on Assessment of Ambient Air Pollutants in Vicinity of Gujarat Composite Limited, Adhmedabad*, National Institute of Occupational Health, Meghani Nagar, Ahmedabad

Normark, A. (2006) 'Appeal by the Building and Woodworkers International (BWI)' in *Chrysotile Asbestos: Hazardous to Humans, Deadly to the Rotterdam Convention*, Building and Woodworkers International and International Ban Asbestos Secretariat, London

Northern Cape State of the Environment Report (2005) *Human Settlements Specialist Report*, prepared for the Department of Tourism, Environment and Conservation by E. Muller and M. Wilkinson, CSIR Environmentek, www.deat.gov.za/soer/ncape/files/human_settlements_specialist_report_5_of_7.pdf, accessed 1 July 2010

NUM (National Union of Mineworkers) (2007) *Submission on Asbestos and Safety, 21 June 2007, Cape Town*, submission made to the Chrysotile Joint Technical Task Force Team, Cape Town, 20–21 June 2007

Nurse, G. T. (1975) 'The origins of the Northern Cape Griqua', *Institute for the Study of Man Paper*, vol 34, pp2–21

O'Neill, R. (2005) 'Burying the evidence', *Hazards*, no 92, pp 4–5, www.hazards.org/cancer/report.htm, accessed 29 June 2010

O'Regan, S., Tyers, C., Hill, D., Gordon-Dseagu, V. and Rick, J. (2007) *Taking Risks with Asbestos: What Influences the Behaviour of Maintenance Workers?*, HSE Research Report RR558, May 2007

Palladino, P. (2002) 'Between knowledge and practice: On medical professionals, patients, and the making of the genetics of cancer', *Social Studies of Science*, vol 32, no 1, pp137–165

Pancholi, T. (1997) 'Comments of the expert on the report submitted by NIOH', unpublished report for civil application No 8617 of 1997, in the matter of environmental pollution with respect to Shree Digvijay Cement Company Ltd

Pearson, V. A. H. and Sims, P. A. (1992) 'An asbestos hazard in north Devon', *Journal of Public Health Medicine*, vol 14, no 4, pp376–379

Peto, J. (1989) 'Fibre carcinogenesis and environmental hazards', in J. Bignon, J. Peto and R. Saracci (eds) *Non-Occupational Exposure to Mineral Fibres*, World Health Organization and International Agency for Research on Cancer, IARC Scientific Publications, no 90

Peto, J., Hodgson, J. T., Matthews, F. E. and Jones, J. R. (1995) 'Continuing increase in mesothelioma mortality in Britain', *Lancet*, vol 345, pp535–539

Phillips, J. I., Renton, K., Murray, J., Garton, E., Tylee, B. and Rees, D. (2007) 'Asbestos in and around Soweto dwellings with asbestos cement roofs', *Occupational Health Southern Africa*, vol 13, no 6, pp3–7

Pollitt, N. (1982) 'Licence to kill renewed', *New Statesman*, 27 August, p6–7

President's Commission (1983) *Report of the Constitutional Committee of the President's Council on the Needs and Demands of the Griqua*, Government Printer, Cape Town

PRIA (Society for Participatory Research in Asia) (2004) *Occupational Health in India: Making a Difference*, Tughlakabad Institutional Area, New Delhi

Price, B. and Ware, A. (2008) 'Asbestos exposure and disease trends in the 20th and 21st centuries', in John E. Craighead (ed) *Asbestos and Its Diseases,* Oxford University Press, Oxford, pp375–396

Price, R. (1998) 'Reversing the gun sights: Transnational civil society targets land mines', *International Organization*, vol 52, no 3, pp613–644

Rake, C., Gilham, C., Hatch, J., Darnton, A., Hodgson, J. and Peto, J. (2009) 'Occupational, domestic and environmental mesothelioma risks in the British population: A case-control study', *British Journal of Cancer*, pp1–9

Raman, Q. (2008) 'Health hazards due to asbestos exposure in India', in D. Allen and L. Kazan-Allen (eds) *India's Asbestos Time Bomb*, International Ban Asbestos Secretariat, London

Rahmanathan, A. L. and Subramanian, V. (2001) 'Present status of asbestos mining and related health problems in India', *Industrial Health*, vol 39, pp309–315

Randeree, A. (1998) *Asbestos Pollution: What Needs to be Done*, unpublished report by the Department of Environmental Affairs, Developmental Social Welfare and Health

Rantanen J. (1997) 'Global asbestos epidemic: Is it over?', in A. Tossavainen, M. S. Huuskonen and J. Rantanen (eds) *Proceedings of an International Expert Meeting on Asbestos, Asbestosis, and Cancer*, Finnish Institute of Occupational Health, Helsinki

Rapp, R. (1999) *Testing Women, Testing the Fetus: The Social Impact of Amniocentesis in America*, Routledge, London

Reid, G., Kielkowski, D., Steyn, S. D. and Botha, K. (1990) 'Mortality of an asbestos-exposed birth cohort', *South African Medical Journal*, vol 78, pp584–587

Reinold Noyes, C. (1940) 'The law and scientific method', *Political Science Quarterly*, vol 55, no 4, pp496–521

Richards, P. (1989) 'Agriculture as performance', in R. Chambers, A. Pacey and L. A. Thrupp (eds) *Farmer First: Farmer Innovation and Agricultural Research*, Intermediate Technology Publications, London

Richards, P. (2006) 'Against discursive participation: Authority and performance in African rural technology development', paper presented at the EASST conference, Lausanne, Switzerland, August 2006

Roberts, B. (2009) 'Gordon Brown accused of u-turn over compensation for asbestos victims', *Mirror*, 10 July, http://blogs.mirror.co.uk/asbestos-campaign/2009/07/gordon-brown-accused-of-u-turn.html, accessed 12 July 2010

Roberts, J. (2002) *The Historical Failure of the Surveillance of Asbestos-Related Diseases in South Africa: Epidemiological Problems and Recent Advances in Detection*, www.asbestosclaims.co.za/docs%5Cthehistoricalfailureofmedicalsurveillance.pdf, accessed 24 August 2007

Rosenau, J. (2002) 'Governance in the twenty-first century', in R. Wilkinson (ed) *The Global Governance Reader*, Routledge, London

Ross, R. (1976) *Adam Kok's Griqua*, Cambridge University Press, Cambridge

Rosser, S. (2000) *Women, Science and Society: The Crucial Union*, Teachers' College Press, New York, NY

Rothwell v Chemical Insulating Co. Ltd and another [2006] EWCA Civ 27

Rudd, R. (2002) 'Benign pleural disease', in D. Hendrick, W. Beckett, P. S. Burge and A. Churg (eds) *Occupational Disorders of the Lung: Recognition, Management and Prevention*, Saunders Ltd, London

Rudd, R. (2004) 'Pleural Plaques', report prepared for the court in *Grieves and others v F. T. Everard & Sons and British Uralite Plc and others* [2005] EWHC 88 (QB)

Ruff, K. (2008) 'Current update on situation at Rotterdam convention meeting in Rome!', coordinator of the Rotterdam Convention Alliance (ROCA), unpublished email communication from IBAS, 29 October 2008

Saaiman, P. (2008) 'Budget Speech' address by P. W. Saaiman, MEC for Economic Affairs, Tourism, Environment and Conservation in the Northern Cape Legislature, Department of Tourism, Environment and Conservation, 12 June 2008, www.info.gov.za/speeches/2008/08062712451001.htm, accessed 12 July 2010

Sarde, S. R. (2009) 'Organizing migrant ship breaking workers in India: A unique experience', paper presented at the Asian Asbestos Conference, Hong Kong, 26–27 April 2009

Scholte, J. A. (2000) *Globalisation: A Critical Introduction*, Macmillan, London

Scholte, J. A. (2002) 'Civil society and democracy in global governance', in R. Wilkinson (ed) *The Global Governance Reader*, Routledge, London

Scoones, I., Leach, M., Smith, A., Stagl, S., Stirling, A. and Thompson, J. (2007) *Dynamic Systems and the Challenge of Sustainability*, STEPS Working Paper 1, STEPS Centre, Brighton

Selikoff, I. J. (1977) 'Disease prevention in asbestos insulation work', *Occupational Safety and Health Series*, vol 27, pp13–26

Selikoff, I. J. and Hammond, E. C. (1975) 'Multiple risk factors in environmental cancer', in J. F. Fraumeni Jr (ed) *Persons at High Risk of Cancer: An Approach to Cancer Etiology and Control*, Academic Press, New York, NY

Selikoff, I. J., Churg, J. and Hammond, E. C. (1964) 'Asbestos exposure and neoplasia', *JAMA: Journal of the American Medical Association*, vol 188, pp22–26

Selikoff, I. J., Churg, J. and Hammond, E. C. (1968) 'Asbestos exposure, smoking and neoplasia', *JAMA: Journal of the American Medical Association*, vol 204, pp104–110

Sethi, N. (2007) 'Asbestos lobby to study health hazards', *Times of India*, 21 June, http://timesofindia.indiatimes.com/home/science/Asbestos-lobby-to-study-health-hazards/articleshow/2137397.cms, accessed 14 July 2010

Shabangu, S. (2001) Budget Vote Speech 2001/2001 by Deputy Minister of Minerals and Energy, National Assembly 8 May 2001, www.info.gov.za/speeches/2001/0105101045a1002.htm, accessed 1 July 2010

Shah, E. (2009) 'Debating knowledge: New spaces of conversation', *Seminar 597 Knowledge in Question*, May 2009, pp69–81

Sharma, S. (1998) 'Transcending boundaries: Occupational and environmental health scenario in South Asia', *Occupational and Environmental Health*, vol 4, no 4, pp1–3

Sikkink, K. (1998) 'Transnational politics, international relations theory and human rights', *Political Science and Politics*, vol 31, no 3, pp516–523

Smit, P. (2009) 'Asbestos banned, but legacy continues', *Engineering News Online*, 30 October, www.engineeringnews.co.za/article/respiratory-asbestos-holds-significant-health-risks-2009-10-30, accessed 1 July 2010

Steele, J. and Wikeley, N. (1997) 'Cases: Dust on the streets and liability for environmental cancers', *Modern Law Review Limited*, vol 60, no 2, pp265–276

Steinberg, H. (2006) 'Pleural plaques: An update', Personal Injuries Bar Association, www.piba.org.uk/assets/docs8/pplecture_PIBA_draft4_170106.doc, accessed 8 August 2006

Stephens, S. (2002) 'Bounding uncertainty: The post-Chernobyl culture of radiation protection experts', in S. M. Hoffman and A. Oliver-Smith (eds) *Catastrophe and Culture: The Anthropology of Disaster*, James Currey, Oxford

Stirling, A. (2009) 'Risk, uncertainty and power', *Seminar 597: Knowledge in Question*, May 2009, pp33–40

Straif, K., Benbrahim-Tallaa, L., Baan, R., Grosse, Y., Secretan, B., El Ghissassi, F., Bouvard, V., Guha, N., Freeman, C., Galichet, L. and Cogliano, V. on behalf of the World Health Organization (2009) 'Special report: Policy. A review of human carcinogens – Part C: metals, arsenic, dusts, and fibres', *Lancet*, vol 10, pp453–454

Terracini, B. (2006) 'The scientific basis of a total asbestos ban', *La Medicina del Lavoro*, vol 97, no 2, pp383–392

Thomas, H. F., Benjamin, I. T., Elwood, P. C. and Sweetnam, P. M. (1982) 'Further follow-up study of workers from an asbestos cement factory', *British Journal of Industrial Medicine*, vol 39, no 3, pp273–276

Thompson's Solicitors (2003) 'Litigation in South Africa against Gencor, Gefco and Msauli: A briefing document for trade unions', 20 March 2003, www.sni.co.za/Gencor%20Brief%2020.03.03.doc, accessed 22 December 2007

Thompson's Solicitors (2007) 'Mesothelioma, the UK's third fastest increasing cancer', 10 August, www.thompsons.law.co.uk/ntext/mesothelioma-uk-third-fastest-increasing-cancer.htm, accessed 1 July 2010

Tiger, L. (1984/1969) *Men in Groups*, Boyars, New York, NY

Touraine, A. (1988) *Return of the Actor: Social Theory in Post Industrial Society*, translated from French by M. Godzich, by University of Minneapolis Press, Minneapolis, MA

Toxics Link (2002) *Fibres of Subterfuge*, July, www.indiatogether.org/environment/articles/asbtos1.htm, accessed 1 July 2010

Trade Union Advisory Committee to the OECD (Organisation for Economic Co-operation and Development) (2005) (Draft) *OHSE Sust/Dev Country Profiles for Asbestos*, 30 May 2005, www.global-unions.org/pdf/ohsewpL_6.EN.pdf, accessed 16 June 2005

Traweek, S. (1988) *Beamtimes and Lifetimes: The World of High Energy Physicists*, Harvard University Press, Cambridge, MA

Turkle, S. (2007) *Evocative Objects: Things We Think With*, MIT Press, Cambridge, MA

Tweedale, G. (2000) *Magic Mineral to Killer Dust*, Oxford University Press, Oxford

Tweedale, G. (2008) 'Asbestos Multinationals in India: The Experience of Turner & Newall', in D. Allen and L. Kazan-Allen (eds) *India's Asbestos Time Bomb*, International Ban Asbestos Secretariat, London

Vail, L. (ed) (1989) *Creation of Tribalism in South Africa*, James Currey, Oxford

van Beusekom, M. and Hodgson, D. (2000) 'Lessons leant? Development experiences in the late colonial period', *Journal of African History*, vol 41, no 2, pp29–33

van Steenbergen, B. (1994) *The Condition of Citizenship*, Sage, London

van Zyl, P. and Gibson, B. (1987) '"Grasping the nettle": A review of the asbestos cement industry's response to the asbestos and health issue 1965–1987', unpublished paper presented at the International Conference on Air Pollution, South Africa

Venkatesan, V. (2007) 'The dilution of a principle', *Frontline*, vol 24, no 22, www.hinduonnet.com/fline/fl2422/stories/20071116505108000.htm, accessed 12 July 2010

Venter, I. (2004) 'SA makes progress in asbestos-mine rehabilitation', *Creamer Media's Mining Weekly online*, 30 July, www.miningweekly.com/article/sa-makes-progress-in-asbestosmine-rehabilitation-2004-07-30, accessed 1 July 2010

Visagie, N. (2008) 'Cabinet to get report on pollution in N. Cape', *Diamond Fields Advertiser*, 24 January

Visvanathan, S. (2009) 'Knowledge, justice and democracy', in M. Reynolds, C. Blackmore and M. J. Smith (eds) *The Environmental Responsibility Reader*, Open University, Zed Books, London

Waldman, L. (2003) 'Houses and the ritual construction of gendered homes', *Journal of the Royal Anthropological Institute*, vol 9, no 4, pp657–679

Waldman, L. (2006a) 'Klaar Gesnap as Kleurling: The attempted making and remaking of the Griqua people', *African Studies*, vol 65, no 2, pp175–200

Waldman, L. (2006b) 'Community, family and intimate relationships: An exploration of domestic violence in Griquatown, South Africa', *Anthropology Southern Africa*, vol 29, nos 3 and 4, pp84–95

Waldman, L. (2007) 'When social movements bypass the poor: Asbestos pollution, international litigation and Griqua cultural identity', *Journal of Southern African Studies*, vol 33, no 4, pp577–600

Waldman, L. and Maat, H. (2007) 'Introduction: Understanding participation through science and technology', *IDS Bulletin*, vol 38, no 5, pp1–5

Waldman, L. and Williams, H. (2009) *As Safe as Houses? Dealing with Asbestos in Social Housing*, a report for the Union of Construction, Allied Trades and Technicians, June 2009

Wang, X. (2009) 'Current situation of asbestos-related problems in China', paper presented at the Asian Asbestos Conference, Hong Kong, 26–27 April 2009

Ward, H. (2002) 'Corporate accountability in search of a treaty: Some insights from foreign direct liability', *Royal Institute of International Affairs Briefing* Paper, vol 4, pp1–11

Warner, J. H. (1995) 'The history of science and the sciences of medicine', *Osiris*, 2nd Series, vol 10, pp164–193

Watts, M. (2005) 'Colonialism, history of', in T. Forsythe (ed) *Encyclopedia of International Development*, Routledge, London

Weiss, W. (1977) 'Mortality of a cohort exposed to chrysotile asbestos', *Journal of Occupational Medicine*, vol 19, no 11, pp737–740

Welch, L. S. (2007) 'Asbestos exposure causes mesothelioma, but not this asbestos exposure: An Amicus brief to the Michigan Supreme Court', *International Journal of Occupational and Environmental Health*, vol 13, no 3, pp318–327

WHO (World Health Organization) (2006) *Elimination of Asbestos-Related Diseases*, http://whqlibdoc.who.int/hq/2006/WHO_SDE_OEH_06.03_eng.pdf, accessed 1 July 2010

Wikeley, N. (1997) 'Turner & Newall: Early organizational responses to litigation risk', *Journal of Law and Society*, vol 24, no 2, pp252–275

Wikeley, N. (2000) 'The victim and the law: Problems in social security adjudication', in T. Gorman (ed) *Clydebank: Asbestos, the Unwanted Legacy*, Clydebank Asbestos Partnership, Clydeside Press, Glasgow

Williams, J. J. (2004) 'Citizenship, community participation and social change: The case of area coordinating teams in Cape Town, South Africa', *IDS Bulletin*, vol 35, no 2, pp19–25

Wilson, R. (2008) *Don't Be Fooled Again: The Sceptic's Guide to Life*, Icon Books Ltd, Cambridge

Wilson, R. (2009) 'Don't get fooled again: Yet more false and misleading claims on asbestos from the Sunday Telegraph', a book blog by R. Wilson, http://richardwilsonauthor.wordpress.com/2009/10/04, accessed 5 January 2010

Wynne, B. (1995) 'The public understanding of science', in S. Jasanoff, G. E. Markle, J. C. Peterson and T. Pinch (eds) *Handbook of Science and Technology Studies*, Sage Publications, Thousand Oaks, CA

Wynne, B. (1996) 'May the sheep safely graze? A reflexive view of the expert-lay knowledge divide', in S. Lash, B. Szerszynski and B. Wynne (eds) *Risk, Environment and Modernity: Towards a New Ecology*, Sage, London

Yarborough, C. M. (2006) 'Chrysotile asbestos and mesothelioma', *Critical Reviews in Toxicology*, vol 36, no 2, pp165–187

Yuval-Davis, N. (1999) 'The "multi-layered citizen": Citizenship in the age of "globalization"', *International Feminist Journal of Politics*, no 1, vol 1, pp119–136

Zaidi, A. (2006) 'Environment: Killer mineral', *Frontline*, vol 23, no 20, pp100–103

Zavestoski, S., Agnello, K., Mignano, F. and Darroch, F. (2004) 'Issue framing and citizen apathy toward local environmental contamination', *Sociological Forum*, vol 19, no 2, pp255–283

Index

Note: page numbers followed by n refer to end of chapter notes

activists
  aims 182
  alternative framings in India 164–167
  challenging dominant framings 159,
      163–164, 170
  grassroots context 91, 94–101
  mobilization 86–88, 102–103,
      159–160, 179
  transnational context 88–94
  understanding of asbestos 81
  see also CPAA (Concerned People
      Against Asbestos); participation
actor-orientated approaches 12, 15
adaptations 68
Alang shipyards 93, 107n
Alimta (pemetrexed disodium) 5
ambiguity 159, 166, 173
  economic imperatives 169
  engineers 34
  UK alternative framings 161, 162,
      163
amosite 4, 80
  banning of in UK 21, 145
  mines 111
  products 112
anthropology 8–12, 126
  see also medical anthropology
Appadurai, A. 75, 87, 102
apprehension 62
ARDs (asbestos-related diseases)
  association with smoking 115–116
  awareness of in UK 21
  case studies 26–29, 30–31

categorizations of 5, 20, 26–27, 29,
    30, 117–118
definitions 25–26, 84, 118, 180
extent of problem 5–6, 104n,
    114–115, 140n, 166, 176n
hidden in India 84–85, 86, 104, 164
impact on identity 8–9, 126–130
  see also masculinity
impacts on family 38–40, 62–63
laggers' attitudes to diagnosis 26,
    28–33, 119, 133–134
latency period 5, 65–66, 79, 117
medical and legal views of 20, 22–23,
    65–66, 117–123
medical screening 57–58
perceptions of North Cape residents
    64–65, 66–68, 115, 116–117,
    119
treating and fighting 5, 123–126
asbestos
  banning of 8, 21, 51, 70n, 113–114,
      145–146, 152–153
  concealment of risk 6, 7
  indirect exposure to 128, 137
  nature and uses of 1, 4–5, 7,
      111–112, 185
  profitability of 32
Asbestos Forum (Northern Cape) 56, 61
asbestos industry
  changing attitudes 153–154, 167
  framings of asbestos 144, 145, 151,
      162–163
  India 74, 88, 89, 90, 95–96

requirement for Indian evidence
83–86, 90, 103
success of 32
see also Cape plc; government/industry
collusion
Asbestos Industry Regulations (1931 UK)
21, 113
asbestosis 5, 65, 119
diagnosis in India 84, 86
legal categorization 20, 22, 25–26,
66, 118
asbestos mines 1, 4, 50–53
closure 8, 70n, 114
communities 128–129, 132–133,
155–156
environmental rehabilitation 58–59,
155–156
female workers 135–136
Zimbabwe 167, 168
asbestos products 59–60, 70n
see also buildings and *in situ* asbestos
Asbestos Regulations (SA) 153, 154
Asbestos Relief Trust (ART) 155, 175n
asbestos substitutes 168–169, 177n
asbestos waste
local use of 59–60
shipbreaking 92–93, 106n, 165
see also pollution
Asbestos Watchdog 162–163, 176n
Asian Asbestos Conferences 165–167

Ban Asbestos Network India (BANI) 75,
88–94, 97–98, 101–102, 103
alternative framings 164
building networks 181–182, 184
Barking and Dagenham 8, 21, 37, 112,
123
emic interpretations of illness 124–
125
exposure of women to asbestos
136–137
see also laggers
Basel Convention 104n, 107n
Blue Lady 92–93

Booker, Christopher 163
Bridle, John 163
Building and Other Construction
Workers Welfare Cess Act (India) 77
buildings and *in situ* asbestos 146–147
South Africa 59–60, 154–155, 156
UK 161–163
Burdett, G. 146

Canada 74
Cape plc 1, 110, 111–112, 136
asbestos products 70n
case brought by CPAA 53, 54, 114
North Cape mines 51, 52
paternalism 37, 129–130
Castleman, Barry 91
chrysotile 4
banning of in UK 21, 145
chemical and structural alteration
162–163
NIOH research 78–79, 150–151
perceptions of risk 80–83, 116, 151
Zimbabwe production of 167–169
Chrysotile Institute 79, 80, 81–82
Chrysotile Joint Technical Task Force
Team 167–169
citizenship 87, 94
transnational 74–75
see also participation
Clayson, H. 122
Collegium Ramazzini 81, 82, 105n
compensation 3
claims in South Africa 8, 53, 69n,
114, 175n
difficulties in claiming in India 77–78,
85, 86
entitlements in South Africa 66, 118,
120, 133, 139
laggers claims on diagnosis 42–43,
134
legal framework of ARDs in UK
19–20, 22–27, 141n
unwarranted claims 20, 148–149, 150
wives ambivalence towards 122, 137

Congress of South African Trade Unions (COSATU) 153
Connell, R. 33
Control of Asbestos Regulations 146
CPAA (Concerned People Against Asbestos) 53–55, 56, 70*n*, 113–114
    asbestos-impregnated construction materials 59–60
    data collection 57–58
    environmental rehabilitation 58–59
crocidolite (blue asbestos) 4, 80, 116
    banning of in UK 21, 51, 145
    mines 111

Dagenham *see* Barking and Dagenham
Damages (Asbestos-Related Conditions) (Scotland) Bill 161
death
    families experience of 38–40, 62–63, 128
    work-related rates 6
Defective Premises Act 146
definitions of ARDs 25–26, 84, 118, 180
denial of hazard
    India 9, 73
    South Africa 51
    UK 162–163
Department of Minerals and Energy (DME) 58
developing countries, target for producers 6–7, 74
Directorate General, Factory Advice, Service and Labour Institute (DGFASLI) 78
disease *see* ARDs (asbestos-related diseases)
Douglas, M. and Wildavsky, A.B. 49–50
Dunne, R.M. 27–28, 42, 43–44, 120
du Plessis, Hein 63

economic imperatives 168–169
Eco-Rehabilitation 59
Edwards, J. et al 10

Egilman, D.S. 80
emic understandings 115–117, 119, 120–123
    alternative framings 163
    North Cape 64–65, 66–68, 69, 123–126
Employees' State Insurance (ESI) (India) 77–78, 85, 86
engineers 34, 41, 47*n*
    laggers as 36, 135
Environmental Public Hearings (EPHs) 151
environmental rehabilitation
    airborne pollution 60–62
    asbestos-impregnated construction materials 59–60, 162–163
    mine workings 58–59, 155–156
Escobar, A. 94
European Union (EU) 35–36, 70*n*
Everest asbestos factory 95–96, 151
Everite 59–60, 153–154, 168–169, 177*n*
expert witnesses 28–29

Factories Act (India) 77
family structure 37–38
    and attachment to home town 65
    Griqua 131
    impact of ARDs 38–40, 62–63, 128–129
Faulkner, W. 34, 41
Felstiner, W. and Dingwall, R. 26
fibrosis 5, 65
fluid on the lungs 5, 65
foreign workers 35–36
Fox, J. 74–75, 92
framings 13, 17–18*n*, 143–145, 169–170
    alternatives for India 164–167, 171
    alternatives for South Africa 167–169, 171
    alternatives for the UK 159–164, 171
    asbestos dangerous but controllable 144, 145–147
    asbestos hazard stressed 152–156

asbestos risks non-existent 150–152
domination of science and risk
  156–158, 172–173, 181
status of pleural plaques 147–150
undeserving citizens 20, 148–149,
  180
Furedi, F. 50

gender 126–130
women's identities 126–127, 132,
  135–137, 138
see also masculinity
Global Asbestos Congresses 165
global civil society and identity 74–75,
  87–88, 91–94
globalization 11, 74, 87
India 76
isolation of the poor 75–76, 94
GMB union (UK)
foreign workers 35
members 2, 20–21, 22, 40–41, 170
Good, B.J. 28, 44
governance 3, 87
through science and technology 170,
  172, 180–181
government/industry collusion
India 79, 96–97, 151
perception of laggers 32, 148
South Africa 51, 52–53, 57, 113, 153
governments
and activism in India 88–89
symbolic nature of monitoring 97,
  99–100, 101
see also government/industry collusion
grassroots mobilization 87, 91, 94,
  97–101
Asia 165–166
*Grieves and others v F. T. Everard & Sons
  and British Uralite Plc and others*
  23–24, 31
Griqua identity 131–133, 135–136,
  141–142n
Griquatown, South Africa 51, 52, 60,
  64, 65, 133

Gujarat, India 95–101, 151, 167
Gupta, A. and Ferguson, J. 11–12, 14

Harding, S. 12–13
hazardous zones, India 86
health insurance, India 77–78
Health and Safety Executive (UK)
  146–147, 161–163
health and safety regulation
India 76–77, 85, 86, 90
South Africa 152–153, 154
UK 21, 146–147, 161–163
healthy worker effect 85, 106n
Held, D. and McGrew, A. 87
Helman, C. 22
Henwood, K. et al 126, 127, 136
High Court of England and Wales 23,
  31
Holland, Justice 24
Hong Kong Declaration Towards a
  Complete Ban on all forms of
  Asbestos 166
House of Lords ruling on pleural plaques
  19–20, 149, 160, 185

IBAS *see* International Ban Asbestos
  Secretariat (IBAS)
identity 3, 11–12, 138, 181
activists 87, 92, 102
of ARD victims denied in India 85,
  86, 104, 164, 181
ethnic 131–132, 141–142n
impact of ARDs on 8–9, 126–130,
  182
women 126–127, 132, 135–137, 138
see also masculinity
IIAC (Industrial Injuries Advisory
  Council) 147–148, 149
ILO *see* International Labour
  Organization (ILO)
India 7, 8, 180
activist mobilization 88–94, 179, 184
alternative framings 164–167, 171
asbestos industry 74, 95–96

denial of hazard 9, 73, 150–152, 157, 158

dominant framing of asbestos 150–152, 157, 158, 172

enhanced differences between workers and activists 75–76

flouting of regulations 95–97

governance and regulation 76–79, 102

grassroots mobilization 97–101

knowledge politics and the asbestos debate 83–86

Industrial Health Research Group (University of Capetown) 55, 57

Industrial Injuries Advisory Council (IIAC) 147–148, 149

Industrial Injuries Disablement Board 148

information dissemination 81, 93, 105–106n

in situ asbestos *see* buildings and *in situ* asbestos

insurance companies, compensation for pleural plaques 23, 25, 149–150

International Agency for Research on Cancer (IARC) 83

International Ban Asbestos Secretariat (IBAS) 25, 75, 92

bursaries 167

work-related deaths 6

International Labour Organization (ILO) 82

Asbestos Convention 70n, 168, 169, 176n

International Programme on Chemical Safety (IPCS) 82

internet 81, 91–92, 93, 105n

Italy 126

Jasanoff, S. 170

Joshi, T.K. 85–86, 89, 91, 92

Kasperson, R.E. and Kasperson, J.X. 6, 18n

Kazan-Allen, L. 6, 52

Kleinman, A. 29

knowledge 12–13, 180–181

anthropological conceptualizations of 10–11

apprehension 62

declining status 29

and participation 56, 68–69, 98, 101

prioritized in UK and India 170, 172

and public experience of ARDs 64–68, 115–117, 163, 182

and sustainability 182–183

knowledge-as-action approaches 55

knowledge politics

the global asbestos debate 79–83

the Indian asbestos debate 83–86

Koegas, South Africa 51, 59, 112–113, 128, 140n

Labour government 147, 148, 157, 185

laggers 8, 20–21, 22

activist mobilization 160–161, 170

awareness of risk 116, 117

case studies 26–29, 30–31

changes in profession 35–37

compensation claims 23–25, 42–43, 134

dismissal of psychological factors 31–32, 33

emic interpretations of illness 124–125

family structure 37–38, 129

nature of profession 34–35

pride 35, 129

social impact of ARDs 38–41, 134–135

suspicion of medical diagnosis 26, 28–33, 119, 120–123

see also masculinity

Leach, M. et al 56, 69

legal categorizations of ARDs 20, 26–27, 29, 30, 117–118, 147–149

Lipschutz, R. 103

Longsmith, Martin 44

lung cancer 5, 20, 22, 25, 65, 105n

diagnosis 26, 85
and smoking 115–116
women 135

McCormick, S. 172–173
McCulloch, J. 140*n*
Manvar, Raghunath 100–101, 102
masculinity 33–41, 181
family and legality 42–43
mine workers 129
undermining 44–45, 130–135,
  138–139
medical anthropology 22, 29–30
Medical Bureau for Occupational
  Diseases (MBOD) 53, 54
compensation for pleural plaques 1,
  66, 118, 120
medical facilities 123
medical science 3
alternative framings for India
  164–165
definitions of ARDs 25–26, 84, 118,
  180
diagnosis of ARDs 26–29, 65–66,
  119–123, 136–137
dominance in framings of risk
  156–158, 173
occupational health in India 85–86
perceptions of ARDs 20, 22–23
suspicions of support for capitalist
  interests 26, 31, 32–33, 165
medical screening 57–58, 95
Mellström, U. 34, 35
Menon, Rajiv 31–32, 33, 41
mental stress 30, 31–32, 122–123
mesothelioma 5, 20, 22, 25, 65–66
association with chrysotile 80
burden of medical interventions 122
case studies 28, 30–31, 33
diagnosis in India 85, 86, 164, 176*n*
fear of 30
risk of 44, 104*n*, 116–117
women 135, 136
Michael, M. 62, 63, 68

mining companies
exports to developing countries 74
and research 81
see also Cape plc
Mohanty, R. and Tandon, R. 89
Mollona, M. 37–38
Moore, H. 127
Moore, S. and Lenaghan, J. 136
Mossman, B.T. and Gee, B.L. 5, 65
multi-sited research 14–15
Murlidhar, V. and Kanhere, V. 85
Myers, J. 52

Nath, M. 85
National Asbestos Summit (SA) 152, 167
National Campaign on Dust-Related
  Lung Diseases (NCDRLD) 78,
  83–84
National Institute of Occupational
  Health (NIOH) (India) 84–85, 86,
  88, 91, 99, 102
research into chrysotile 78–79, 150–151
National Institute for Occupational Safety
  and Health (NIOSH) (US) 82
National Lung Campaign (India) 90–91
national policy processes 143–145
dominant framings in UK 145–150
dominant Indian position 150–152
domination of science and risk
  156–158
South African dominant framings
  152–156
through science and technology 170,
  172
National Union of Teachers (NUT) 162
New Sahyadri Industries Ltd (NSIL) 151
NIOH see National Institute of
  Occupational Health (NIOH)
  (India)
North Cape residents
fighting disease 123–124, 125
identity 128–129, 130–133
perceptions of risk 64–65, 66–68,
  115, 116–117, 119, 125–126

Northern Cape Executive Council 156
Noyes, R. 26

Occupational Health and Safety
    Association 100–101, 102
Occupational Health and Safety Centre,
    Mumbai 164, 176*n*
open-cast mining 51
ovarian cancer 71*n*

participation 49–50
    concern about visible fibres 60–62
    encouraged by government 56,
        151–152, 170
    engagement with scientific data
        collection 57–58
    environmental rehabilitation 58–59, 60
    grassroots mobilization 87, 91, 97–98,
        101, 165–166
    knowledge-as-action approach 55
    and risk 68–69
    see also activists; Ban Asbestos Network
        India (BANI); CPAA (Concerned
        People Against Asbestos)
Paryavaran Mitra 98–100, 101, 102
patients
    labelling by doctors 22
    suspicion of medical diagnosis 26,
        28–33, 119, 120–123
Peto, J. 146
physician-centred approaches 22, 28
pleural effusions 5, 65, 141*n*
pleural plaques 5, 65
    challenging dominant framings
        160–161
    compensation for in South Africa 1,
        66, 118, 120, 132–133, 139
    compensation for in UK 20, 23–25,
        42–43, 134, 149–150
    dominant framing in UK 147–150
    legal ruling in UK 19–20, 149,
        159–160, 185
    medical view of 22–23, 118, 139
    recognition in South Africa 49, 139

similarity to asbestosis 25–26, 66,
        118, 119
    victim's perspective 26, 30, 44–45,
        120–122, 133–134
pleural thickening 5, 20, 22, 65
politics 3, 8–9, 170, 173
    pathways to sustainability 182–186
pollution
    airborne 60–62
    Italy 126
    South Africa 51–52, 58–59, 114–115
poverty
    India 76, 90
    Northern Cape 8, 52, 69–70*n*
PRIA (Society for Participatory Research
    in Asia) 83, 90–91, 91
Prieska, South Africa 8, 51, 52, 55–56,
        132–133
    airborne pollution 60–61
    environmental risks 60, 156
    establishment of CPAA 53–55,
        113–114
    family and social support 65
    medical screening 57–58
    rehabilitation of mine workings 58–59
    residents' perceptions of risk 64
    risk of ARDs 114
production methods 145
products 1, 4–5, 7, 59–60, 111–112
    phasing out in SA 114, 154–155
protest 3, 6
    see also activists
psychiatric assessments 29

Randeree, A. 57
religion 95
remedies 124
restructuring 47*n*
Richards, G.A. 57
Richards, P. 55
rights-based world views 74–75
risk
    concealment of 6
    control in UK 21, 145

experience of death 62–64
global perceptions of 80–83
in situ asbestos 146–147, 154–155,
    161–163, 174*n*
and public understandings of science
    49–50
understanding through participation
    61, 68, 69
workers' and residents' perception of
    64–68, 115–117
see also framings
Rotterdam Convention 82, 83, 84, 105*n*
Prior Informed Consent (PIC) list 73,
    78–79
rounded atelectasis 5, 65
Rowan, Paul, MP 162
Rudd, R. 22–23, 118, 119

safe levels of exposure 105*n*, 174*n*
in situ asbestos 146
South Africa 52
UK 21, 113
Scholte, J.A. 88, 89, 90
school buildings 162
science 45*n*, 179
accountability 182
and anthropology 9–11
politicalization of 183–184
role of 12–13, 26, 180
and socio-cultural values 68–69
see also medical science
Scotland, diagnosis of ARDs 26
Scottish Parliament 161
shipbreaking 92–93, 106*n*, 165
Shree Digvijay Cement Company 98–99
smoking 115–116, 120
social class 161
social and environmental framing 159,
    173
UK 161, 162, 163
see also South Africa
social movements 3
South Africa 4, 7, 111, 179, 181
alternative framings 171

banning of asbestos 8, 113–114,
    152–153
creating political awareness 53–55
dominant framings of asbestos
    152–156, 157, 158, 172
environmental rehabilitation 58–62
ethnic identity 130–133
families' experience of death 62–64
living with asbestos pollution 64–68,
    125–126
medical screening 57–58
pathways to sustainability 183–184
poverty in Northern Cape 69–70*n*,
    112–113
recognition of ARDs 1, 49
rural participation 55–56
state pensions 66, 118, 133
Zimbabwean opposition to ban
    167–169
see also asbestos mines; Medical Bureau
    for Occupational Diseases
    (MBOD)
Steele, J. and Wikeley, N. 45–46*n*
subcontracting 35–36
sustainability and social justice 13,
    182–183
pathways to 3–4, 183–186
symbolism and meaning 2, 3

Tata Memorial Hospital 164, 176*n*
Thermal Insulation Contractors
    Association (TICA) 36–37
trade in asbestos 74, 77, 167
trade unions 153, 166
UK 159–160, 161
see also GMB union (UK)
transnational citizenship 74–75, 87–88
mobilization in India 88–94,
    102–103
tuberculosis (TB) 120
Tweedale, G. 46*n*, 161, 180

unemployment 52, 70*n*, 76, 90,
    130–131, 132

Union of Construction, Allied Trades and
    Technicians (UCATT) 161
United Kingdom (UK)
  alternative framings 160–164, 171
  banning of asbestos 8, 21, 51, 113,
      145, 184–185
  colonialism and Empire 7–8, 51, 111
  dominant framings of asbestos
      145–150, 157, 158, 170
  factory closures 51
  health and safety regulation 21,
      146–147, 161–163
  legal ruling on pleural plaques 19–20,
      22–25, 149, 159–160, 185
  mobilization of opinion against
      asbestos 159–160, 179
  persistence of ARDS 114–115
  see also laggers

values 26
van Steenbergen, B. 94

welfare systems 134
wet processing 95
Wikeley, N. 26, 119
women and children 51
    see also gender
World Health Organization (WHO)
    82–83
World Social Security Forum 174*n*
Wynne, B. 50, 63

Yuval-Davis, N. 75, 102

Zaidi, A. 164
Zimbabwe 167–169